MW01629297

BARNSTORMING

In Eastern Pennsylvania & Beyond

Jeffrey L. Marshall

Edited by Patrick J. Donmoyer

BARNSTORMING

In Eastern Pennsylvania & Beyond

Jeffrey L. Marshall

Edited by Patrick J. Donmoyer

Pennsylvania German Cultural Heritage Center
Kutztown University of Pennsylvania

2021

Barnstorming in Eastern Pennsylvania & Beyond
By Jeffrey L. Marshall
Edited by Patrick J. Donmoyer

Volume IX of the Annual Publication Series
Pennsylvania German Cultural Heritage Center
Kutztown University of Pennsylvania

Printed in the United States on acid-free paper by Masthof Press, Morgantown, PA

All photography by Jeffrey L. Marshall unless otherwise stated.
Supplemental photography by Patrick J. Donmoyer on pages 6, 7, 8, 77, 133, 169.

Text and photography copyright © 2021 Jeffrey L. Marshall

Layout and design copyright © 2021
Pennsylvania German Cultural Heritage Center, Kutztown University

All rights reserved. No part of this book may be reprinted or reproduced or utilized in any form or by any electronic, mechanical or other means, now known or hereafter invented, including photocopying and recording or in any information storage or retrieval system, without permission in writing from the author and the publisher.

ISBN: 978-0-9987074-6-4
Library of Congress Control Number: 2021915185

22 Luckenbill Road
Kutztown, Pennsylvania 19530
(610) 683-1589
heritage@kutztown.edu
www.kutztown.edu/pgchc

Masthof Press
219 Mill Road | Morgantown, PA 19543
www.masthof.com

CONTENTS

Light gently plays upon the interior of a stone barn in Lancaster County.
Opposite: A classic decorated barn in Virginville, Berks County.

FOREWORD

Barns are iconic features of the American landscape. Standing as monuments to earlier times when a majority of Americans lived or worked on farms, barns capture the spirit of agrarian settlement and embody centuries of farming culture. The oldest of these structures occupy a significant place in the American imagination, evoking a sense of reverence and awe for the achievements and hardworking values of early American farmers from all walks of life and backgrounds.

Among the wide variety of American barns, the Pennsylvania barn holds a special place in the agricultural traditions of North America. Combining features found in Alpine Switzerland and the British Isles, this distinctive Pennsylvania form was culled from the agricultural traditions of the most culturally diverse of the thirteen original colonies. As the breadbasket of the Mid-Atlantic, the commonwealth's agricultural legacy was shaped and defined by immigrant and indigenous farming traditions, and a unique Pennsylvania diaspora introduced the Pennsylvania barn to farming communities throughout North America.

A residential farm in Upper Bucks County.

While some of us are fortunate enough to encounter these barns in our daily lives, travels, and work, relatively few people know the intimate experience of entering these veritable sanctuaries of agriculture. Whether a barn is on a working farm or stands retired on a residential property, there is a stillness that permeates the interior. Amidst the lofty heights of the mows and the cool depths of the stables, there is an accumulation of stories in the hewn beams and time-worn floors, left behind by the working generations of farm families and their animal companions.

As character-defining features of the cultural landscape with unique regionalisms and distinctive decorations, barns express a relationship to the land through their site-specific construction materials in locally-sourced timber, stone, and iron—not merely constructed from, but part, of the landscape itself. The very existence of these monumental structures once depended upon a highly cooperative and community-based system of reciprocity between neighbors of all kinds, including humans, animals, plants, and the soil. These complex webs of interconnections place historic barns at the intersection of ecological, cultural, and economic identities in which modern rural and suburban communities continue to seek meaning and balance.

In *Barnstorming in Eastern Pennsylvania & Beyond*, Jeffrey L. Marshall explores this quintessentially American structure through the eyes of a barnstormer, in colorful photo essays and plain language, which will appeal to a broad range of audiences at every level of interest, from the academician to the hobbyist.

Sharing forty years of fieldwork as a regional leader in historic preservation, Jeffrey explores the wide range of experiences, meanings, and discoveries that take place within the barns of the region. He introduces ways to identify and understand unique barn forms and functions, with special emphasis placed upon details of use and significance in farming operations. Balancing historical and regional terminology with contemporary interpretations, Jeffrey navigates the complexity of Pennsylvania's barns while avoiding rigid classification systems and excessive jargon. Whether in distinctive framing patterns or adaptive reuse of old timbers; in marks left behind by farmhands or formal date stone inscriptions; or in ventilators executed in wood, brick, or stone, Jeffrey's intellectual curiosity and collaborative approach to interpreting barn architecture is at once accessible and refreshing.

While serving as president of the National Barn Alliance, Jeffrey was once asked why he was attracted to barns. He responded that they are much more than architecture: "Barns have a heart, a voice, a soul; the barn's heart, voice, and soul are also the heart, voice, and soul of us as a community, nation, and people."

Jeffrey's particular method of encouraging careful observation of architectural significance lays the groundwork for cultivating community and personal connections to the region's barns as cultural icons worthy of preservation. This unique guide will encourage present and future generations to explore and appreciate these historic structures, with the potential to make barnstormers of us all!

Patrick J. Donmoyer
May 17, 2021
Kutztown, Pennsylvania

Slowing down to look at a barn with a car closing in behind me… One of the hazards of barnstorming!

"Too often barns are taken for granted as parts of the scenic background down country roads... every barn reflects in some way the spirit of the age in which it was built with visible comments of the skill of the men who built it... A man's barn bespoke his worth as a man. It expressed his earthly aspirations and symbolized the substance of his legacy to his children."

Berenice M. Ball
Barns of Chester County, Pennsylvania

INTRODUCTION

Barns are fascinating structures. Once the most dominating feature on the rural landscape, most people only get to see old barns traveling along the ever-suburbanizing landscape. Most Americans have probably never physically set foot in one. Certainly, at their core, barns were built to reflect the needs of the farmers who had them constructed. But to many, barns are much more than places to store crops and shelter livestock, more than cow houses and storage buildings. One needs to understand them to truly appreciate them. Barns are like people; they have personalities and flaws. They come in different sizes, shapes, and colors.

Barnstorming is a great word to describe going out and seeing barns across the country. In this sense, the word is reverse engineered from its more common definitions which include conducting a campaign or speaking tour in rural areas by making brief stops in many small towns or its reference to early aviation daredevils.

No matter where the road may lead, passing an old barn is not just passing by weather-beaten walls. The barns still standing in the distance represent the history of this country. The American stories embedded in the interior and exterior walls, recount the tales of our country. When listening to stories told about old barns—some are sad, and others are happy. There are funny stories, and scary stories. Regardless of what kind of stories you prefer, there is always a barn story that will entertain and delight you.[1]

Barn preservationists can argue that a full appreciation of barns can only be gained with an understanding of their function, the purpose they were originally designed to serve. For the rest of the community, barns have a beauty and a whole lot of character that can be appreciated just by taking the opportunity to observe. Barns and family farms have become a cultural icon symbolizing American virtues including self-sufficiency and public-spiritedness, close-knit family and community ties, the compatibility of physical work and mental elevation, and the promise that effort and thrift would lead to abundance. This book is designed to kindle that appreciation and unlock the hidden stories of barns in this region.

The Rolling Green Farm in Solebury Township, Bucks County, is highly visible along the road from Doylestown to the Delaware River. Its location in an open field enhances its architectural merits. The main barn is an English style ground barn. The exact date is not known but the accompanying house has a date of 1748. The offset gable roof section to the left is an unusual feature.
Opposite: A double ground barn with dates of 1758 and 1822 in Plumstead Township, Bucks County

GROUND BARNS

Not all barns in the area are two story bank barns. There are a number of one-level ground barns that often represent eighteenth-century barns. Solebury Township in Bucks County seems to have the greatest number of stone ground barns anywhere in the state. These barns generally have a central threshing bay flanked by bays that were used as stables with lofts above. People have traditionally used the term English Barns to differentiate ground barns from two-level bank barns, but ground barns were built by German-speaking settlers as well. The majority of extant ground barns in Bucks County appear to have been built on properties owned by English speaking Pennsylvanians.

This early form was held in common throughout the region of Southeastern Pennsylvania and was built with a variety of materials and construction methods, including stone, log, frame, and combinations of these techniques. Finding barns that do not fit into standard classifications is part of the life of a barnstormer.

The Moses Paxson Barn in Solebury Township, Bucks County, has a date of 1785 in the gable and the same date in several locations on the interior walls. The detail the mason took on the gable end (below) to align stones of similar appearance, so that the ventilation slits appear to cut through the stones makes this a special barn.

The Heckler Plains barn in Lower Salford Township, Montgomery County, is one of the region's most pristine examples of an early German ground barn (*Grundscheier*) also called a bottom (*Boddem*) barn. The roof is supported by a pair of traditional *liegender Stuhl* trusses.

A two-level English stone ground barn in Solebury Township, Bucks County.

The ground barn on the Dickinson Homestead in Plymouth Meeting, Montgomery County, is another example of an English influenced ground barn. There is evidence of a gable addition that was supported by conical piers.

The Abiah Taylor barn in Chester County is an example of a quintessential English ground barn. It may also be one of the oldest barns in the region.

This form of English ground barn also can be found in New Jersey as seen in this barn located in Mercer County.

FOREBAY BARNS

The part of the barn which is located toward the front (barnyard) wall that extends over the stable wall of the basement, providing a covered, protected area, is called the *forebay*. It is one of the key elements of the Pennsylvania German Schweitzer barn. The advantages of this feature eventually made it a standard element in nineteenth century Pennsylvania barn design. Many nineteenth century texts refer to it as an over shoot.

Some forebays project out above the stable doors while others were enclosed within the main body of the barn. In the former case, called "open forebay" barns, the forebays are either cantilevered or supported by posts or masonry piers.

Above & Opposite: The front of a Schweitzer barn located in North Middleton, Cumberland County, with a projecting forebay providing a sheltered area and protected stable doors from sun in the summer and snow in the winter.

A detailed view showing how the floor joists extend out beyond the stable wall to support the wood frame forebay above.

Above: A Schweitzer style barn in Haycock Township, Bucks County, with its distinctive uneven gable roofline, which is longer on the forebay side than on the bank side.

Right: A built-in forebay ladder on the Jacob Kauffman barn, an early 19th century standard Pennsylvania barn in the Oley Valley, Berks County.

A variation from the projecting, overhanging forebay first appeared in the counties surrounding Philadelphia, where the forebay is formed by moving the lower stable wall several feet back, creating a recessed area below the forebay. Barn historian Robert Ensminger suggested that these areas, with their English traditions, incorporated the benefits of the Germanic forebay, which was so dominant just to the north and west. What has become known as the closed forebay standard barn may represent another example of the fusion of English and Pennsylvania German barn forms. These barns are stone on three sides with a frame wall above the forebay recess. "Their stone construction, balanced gable, and shallow roof slope all conform closely to the typology of the English Lake District bank barns, which were common closer to Philadelphia. The creation of an overhang by recessing the front stable wall and replacing stone with frame for the forebay front wall provided an easy modification to achieve the popular forebay style."[2]

The wooden front forebay wall was supported by corner piers or pillars that helped stabilize the stone end walls. Often these pillars were "L" shaped. Although common on barns built by English and German populations, the Pennsylvania Dutch language term *Peiler Eck* (pier corner) is used to describe the feature.

Occasionally farmers installed a windlass under the forebay to use as a hoist as in the example above in Pike Township, Berks County.

A view of the sheltered area created by the forebay. Note that the sill of the forebay rests on top of the extended floor joists. This photograph demonstrates how one of the main purposes of a forebay was to provide a shaded area as most forebays were oriented to face sunny southeast.

Tapered conical piers are occasionally found on closed forebays as well as extended forebays or straw sheds. A newspaper advertisement for this Solebury Township barn stated "the barn was built in 1835, and is believed to be one of the largest and best constructed buildings of its kind in the county. It has two threshing floors, with granaries and mows on either end."

The barn at Graeme Park near Horsham, Montgomery County, is a large stone structure with a posted extended forebay supported by tapered conical piers more generally associated with barns in Chester County.

Many farmers extended the forebays of their barns for additional storage. The original forebay exterior of this barn in Upper Hanover, Montgomery County, is visible (above) and the original forebay joists are visible projecting out from the right of the lower image. Note that the longer joists are reused timbers. One has a series of empty mortises while a closer one is a reused roof plate with stepped notches for rafters.

Sometimes it is just the fact that it is just a good, solid, working barn like this example in Lower Milford that attracts our appreciation.

Farmers often added projecting extended forebays for additional storage, like this barn in Maxatawny, Berks County.

This example in Franconia Township, Montgomery County, boasts iconic images of farm animals on modest white siding.

Fourteen-pointed stars, a decorative border, arched window, and the painted date of 1850 accent this barn in Upper Milford Township, Lehigh County.

As an alternative to a projecting forebay, this standard barn in Bedminster, Bucks County, features a characteristically recessed stable wall under the frame front wall of the forebay.

The path through the snow to this barn in Solebury Township, Bucks County, reminds us that the farmer can't take a snow day from work.

The snowy late afternoon accentuates this stone barn with the extended forebay located in Chester County, near Morgantown.

Doors open, ladder at the ready, an honest, simple barn in Lower Saucon, Northampton County, evokes a hard working past including its glory days. The barn has pediments above the louvers and a now faded barn star which were signs of pride when constructed.

This forebay view in Durham Township seems to represent the decline of active farming in Bucks County. This barn has been renovated since this picture was taken in 2016.

The hallmark of the Pennsylvania bank barn is a projecting forebay that forms a sheltered area in the barnyard. This Bucks County farmer proudly displays the American flag on his forebay.

What sets this barn apart from similar barns one sees driving through the region is the arched door that pierces the extension of the gable end wall providing access to the area under the forebay, as well as the swallow hole in the gable, the hood above the door and the gambrel roof silo. The barn, located in Earl Township, Berks County, was rebuilt between 1880-1917 after a fire.

This conical pier is located under the front of the original forebay before it was extended on this barn in Hulmeville, Bucks County.

UNDER THE FOREBAY

The classic Pennsylvania barn had a stone stable level to shelter animals while the superstructure above the stable level, whether log, frame, stone or brick, was used to store and process the products of the barn. These areas of the barn generally do not have as much detail as the area above them but still contain character-defining elements.

The term forebay can spark some heated conversation. My colleague Patrick Donmoyer and I have gone back and forth on the topic. Some people are "structural" forebay people and to them the forebay is a projecting structure and others are "spatial" forebay people where the forebay is a sheltered area in the front of a barn and it doesn't matter how it is formed.

If the "bay" were to be the space under the overhang, rather than the structure above, Donmoyer poses the valid question why the forebay is often paired with descriptors like "supported" or "posted" or "cantilevered." These don't seem to make any sense in the situation unless they are paired with the concept of the overhanging structure above. I would think "closed" could also apply to the idea of the overhang.

This barn in Tinicum Township, Bucks County, demonstrates the minimum depth of a forebay—just enough to protect the open door.

I would think that the common usage of the word "bay" in barn terminology—as being a subdivision of the interior space on the upper story demarcated by bents (framing units perpendicular to the ridge)—may hold additional clues about the original intent behind the word forebay.

The traditional use of the terms overhang and forebay interchangeably works for a Schweitzer barn or an extended forebay barn where they are delineated from the main barn but causes murkiness when discussing a Pennsylvania Standard barn without extended side walls or when discussing closed forebay barns. In neither standard barn form is the interior of the portion of the barn above the recess a distinct bay separated from the mows or threshing bays behind it within the barn.

For Schweitzers, likely the earliest of all forebay barns, the forebay was open on the second-floor interior, and usually walkable from one end to the next. Thus, this front "bay" would have earned its name rather aptly by virtue being an open bay at the front of the barn. This idea would be in sync with the German idea of the *Vorbau*.

Buckets and wheelbarrows are part of every horse barn. Horses cooling by the fan in their stall is a newer phenomenon.

Above: One has to look carefully to see the original painted arches on this barn in Nockamixon Township, Bucks County.

Left: Horses show the practicality of Dutch doors in this barn in Plumstead Township, Bucks County.

Above: An ample, unsupported forebay provides shelter for modern equipment on this farm in Centre County.

Right: A wide extended forebay in Lower Milford, Lehigh County. Even nearly three centuries after the forebay was introduced, farmers still benefit from this feature.

Located on the grounds of the elegant property known as "the Highlands" in Whitemarsh, Montgomery County, is a unique barn with a projecting bay on the barnyard side and tall arched windows on the bank side. Construction on the barn reportedly began in 1799-1800.

ONE-OF-A-KIND BARNS

Most barn and farm buildings were built with the simple idea that they were necessary to support the operation of a farm. It seems that many times aesthetics were secondary to functionality. Then there are a number of special barns. They are special because they are outstanding examples of a particular style or construction method, because they are so large, or so highly decorated. The names of the vast majority of the barn architects and builders have been lost, but their work continues to speak to us today.

The cost of many of these barns far outweigh the pure economics of agriculture. They were made as a statement by their owners. The statement may be "look how successful I am" or it may be "I am building this for my children, my grandchildren, and their children. Not enough research has been done to determine whether the economic return on their investment offset the extra cost.

A double-decker barn has two threshing floors, one above the other. This barn is a rare double, double-decker barn in Bucks County. The left section of this double barn in Upper Makefield Township, dates from 1807 and the larger section to the right dates from 1835. The newer section is a double-decker barn. When it was constructed, the older section was raised to a common ridge line. The barn has been adapted with the installation of a number of windows since these photos were taken.

Above: The 1910 Neff Round Barn near Centre Hall, Centre County, rising 56 feet to its cupola is a Pennsylvania gem!

Below: An interior view of the Neff Round Barn. One of the benefits of a round barn with a silo in the center is that it minimized labor associated with feeding cattle.

A unique alternative to a recessed forebay under a frame front wall was a full stone front wall with an arcade to support it. Two other barns in Bucks County and one in Montgomery County have three arches; this is the only one known to have five arches.

When this barn in Doylestown, Bucks County, was advertised in the November 22, 1837 *Bucks County Intelligencer,* it was described as "a stone barn 33 by 55 feet, the front dressed free stone and handsomely raised on five arches".

Most three-level double-decker barns are based on stone English Lake District barns with solid stone front walls. A double-decker closed forebay standard Pennsylvania barn is rare. This barn in Solebury Township in the upper photograph dates from the 1853. It was built for J. S. Williams, Esq. who was the Secretary of the Solebury Farmer's Club. Beautiful cut quoins on the sides of the forebay wall and the exterior corners of this large barn makes it one of the most dramatic in the county. The height of the barn required an extremely tall ramp to get to the threshing floor which on the highest deck level. The ramp ends away from the barn requiring a timber bridge with a bridge house. There are doors for unloading wagons under the bridge.

Below is a double-decker barn in Newlin Township, Chester County, dating from 1820. It is a variation of the stone barns in the English Lake District, which likely influenced these early barns in Pennsylvania.

It is somewhat odd to say that one of most interesting Pennsylvania barns is not in Pennsylvania, but across the Delaware River in Warren County, New Jersey. Many generations of barnstormers have stopped in their tracks when they turn onto a narrow country road and see the Cline (Kline) Barn in all its glory.

It is a large full, double-decker stone barn. If that was not impressive enough, the barn is highlighted by four arches forming a closed forebay on its southerly elevation. The north side has a large stone bridge needed to reach the upper, third, level of the barn. The bridge has a large arch complementing those on the barnyard elevation. Under the arches of the ramp and under the ramp itself is a large root cellar of barrel-vaulted construction. Like early nineteenth century stone barns in Pennsylvania, the walls of this barn are pierced with numerous regularly located louvered ventilators.

Above & opposite: The 1816 Cline Barn ranks high on the list of the "best barns" in the area. It is a double-decker barn with stone arcade on the barnyard side and a tall ramp on the bankside located in Pohatcong Township, Warren County, New Jersey.

Ensminger includes the barn in his book *The Pennsylvania Barn*. He notes that the barn combines the rare stone-arch forebay form with the Quaker double-decker barn form. There is a second, slightly smaller barn of a similar form located a short distance from this barn.

The barn is part of a group of buildings including a large stone house, wagon sheds, corncribs, and a mill. The prosperity of the mill owner is on display in the "one-of-a-kind" barn. Citing Dennis Bertland's *Early Architecture of Warren County, New Jersey*, the barn is dated as circa 1816.[3] "The date is consistent with other stone arch forebay barns in Pennsylvania and also with the stone-column forebay supports of Chester County [Pennsylvania]." While noting masonry arches above doors and windows are found in Pennsylvania German houses and barns, he states "the evidence presented more strongly suggests English origins."[4]

Hidden behind the siding of this Schweitzer barn in Pike Township, Berks County, is a log barn.

BANK BARNS

In addition to the forebay, what distinguishes Pennsylvania barns from other two-story barns is the fact that they were accessible at ground level, on two separate levels. The front (forebay) barnyard was at grade while entry to the threshing floor and flanking mows of the upper story were accessible at the upper elevation. This dual level access was accomplished by building a barn into a hillside or constructing an earthen ramp or bank.

Barn builders occasionally did not mound the "bank" up against the barn wall as hydrostatic pressure would cause the wall to deflect or even collapse over time. An outer wall was constructed a distance from the barn and a ramp or bridge crossed the gap to the exterior barn wall.

Above: A rear view of a double-decker barn in Newtown, Bucks County, showing the steep ramp with a bridge leading to the threshing floor. It was advertised for sale in 1871 as "one of the largest in the township, built of stone, with double floor."

Many nineteenth-century bank barns had frame superstructures on a stone lower level, and nineteenth-century newspaper advertisements refer to these as "frame barns, stone stable high". This view is of the bank side of a barn in Buckingham Township, Bucks County. The bank side of the barn where hay wagons could enter the upper level of a two-story bank barn. The upper level contained a threshing floor flanked by hay mows. Animals were stabled on the basement, or first floor level accessed from the barnyard or gable ends.

Barns change with the seasons and the light.

Proud guardian of the countryside, the barn stands solemnly as a lasting reminder of America's rural heritage. But the barn has begun to disappear from the American landscape. Obsolete for modern farming needs and too expensive to maintain as family heirlooms, old barns appear destined to be preserved only in photographs and memories.

Barns contribute to the landscape and help define the history of the location—how farming was carried out in the past, and how the area has been settled throughout the ages. They can show the agricultural methods, building materials, and skills that were used. Most were built with materials reflecting the local geology of the area.

Barns are working buildings; they are the largest tool on a farm. Like any tool, their shape and size reflects the way in which they are used. A barn's shape, size, and features reflect the job it was intended to do.

As farming practices developed over time, the types of barns that farmers built also changed. Although family farms continue to operate as suppliers for local population centers, the middle of the twentieth century heralded the decline of small farms. Changes in the way American's ate, increasing property values, and the growth of giant agribusinesses meant that family farms had a difficult time making a living.[5]

Equipment and vehicles add a sense of activity at this barn in Alexandria Township, Hunterdon County, New Jersey.

A handsome cupola and a brick stable wall under the forebay make the Stover barn in Tinicum Township, Bucks County, a favorite. The opposite photograph shows the barn winnowing doors open (the railing is a modern addition).

The Stover barn dates from the latter part of the nineteenth century when the property was owned by Jacob Stover. Jacob and his three brothers were third generation millers. The barn is a reflection of a prosperous man. In addition to the milling and lumber business, Jacob Stover was a heavy stockholder in the Alexandria Delaware Bridge company and served as its president.

The barn appears to have been constructed circa 1860 when Jacob Stover turned his attention to agricultural pursuits before he went back into the lumber business with his brother. He continued in the lumber business until 1879. In 1886, he purchased from his brothers, Henry and Jordan H., "The Stover Flour Mills," and manufactured flour and feed until his death.

The barn also contains significant features related to Jacob Stover's lumber business, including numerous timbers in the barn's frame with pairs of "wicket holes"—remnants of the process of floating logs down the Delaware River to the sawmills in Bucks County. These pairs of holes show the location of bent hickory lock-downs or wickets, which were driven into individual logs to secure them to a raft of timber. Similar pairs of holes have been documented in other barns in the area on both sides of the river. Depending on how the raft's logs were sawn, the wicket holes could appear as pairs of oversized circles that are often confused with holes drilled to secure mortises and tenons; made even more confusing when the sawing of a log resulted in only one of the two holes ending up in a particular piece of finished lumber. We often lose sight of the very real connection between natural resources and historic resources. Pairs of wicket holes help us recall the connection.

The Jacob Uhler barn in Riegelsville is one of the grandest of the barns with round ventilators. With the date of 1819, it is also one of the earliest of this subtype.
Opposite: Loopholes on a barn built in 1806 in West Marlborough Township, Chester County.

BARN VENTILATION

Hay that becomes too wet from rain or dew, or that was not allowed to dry sufficiently in the field, will go through a curing process in storage. During the curing process, heat is produced. The storage of hay required ventilation for proper curing without the disastrous consequences of the hay igniting through spontaneous combustion.

Farmers recognized that hay storage structures should have as much ventilation as possible while still protecting the hay from rain and sun. A number of techniques were used to provide adequate ventilation in barns.

The most common form of ventilating stone barns was the vertical slit or loophole. While the term loop, or loophole, was a term used as far back as the fifteenth century to describe castle windows, American barn loopholes were not embrasures designed to allow weapons to be fired out from a fortification while the firer remains under cover.

Top: A barn in West Lampeter Township, Lancaster County, with loopholes and an owl hole in the gable.

Above & Right: Light pours through a loophole in a barn in Solebury Township, Bucks County.

This barn in Nockamixon Township, Bucks County, is one of a number of barns with unusual round ventilators rather than the more typical ventilation slits.

Loopholes were aligned in several rows on all sides of the barn and were flared on the inside. This design may have been used consistently due to its aerodynamic properties that pulled air out and prevented rain from entering. Loopholes continued to be used to ventilate masonry barns into the nineteenth century and were supplanted by fixed wood louvers in frames.

A few barn builders made round brick vents resembling portholes set into stone. We are not aware of any documentation regarding the origin of this feature or research into the extent of the region where they are found. Examples have been documented in Bucks County and Northampton County, Pennsylvania, and across the Delaware River in Hunterdon County and Warren County, New Jersey. Those that are dated range from circa 1820 to 1840.

There is a sub-type of porthole barns that have a single round ventilator at the gable peak which may relate more to an owl hole to allow rodent-eating birds a place to roost rather than holes for the ventilation of hay.

Two views of the Trauger barn of Nockamixon Township, Bucks County, a beautiful barn made even more spectacular by its prominence on a hill surrounded by permanently protected farmland.

The Oberly Barn in Greenwich Township, Warren County, New Jersey, obviously sits right on the road. Approaching the barn from either direction one sees the round ventilators in neat rows.

A swallow hole outlined in white paint at the peak of the roof above the linked white arches of the gable window are some of the highlights of this barn in Washington Township, Berks County. The use of spaced bricks in hourglass patterns offer additional decorative elements. The window trim at the basement level and the brick vents are set proud from the stonework, suggesting that the barn stone work was originally not exposed, but perhaps stuccoed.

Above: The decorated brick gable of the Dillard barn in Franklin County.

Opposite below: This brick barn in New Hanover Township, Montgomery County, features an arched doorway under one end of the closed forebay, providing easier access to the barnyard and stables.

Brick barns are more common in the south-central part of Pennsylvania than in the southeast part of the state. The easternmost examples are located in northwestern Bucks County. Brick barns become more common once you travel west of Harrisburg. Franklin County may have the most examples of this quintessential Pennsylvania type of barn but they are found extensively in the Cumberland Valley. There are more documented in several counties in Maryland adjoining Pennsylvania.

While there are some early examples, most date to circa 1850. Shoemaker notes in *The Pennsylvania Barn* that there were only ten documented brick barns assessed during the 1798 Federal Direct Tax.

Decorative patterns produced by omitted bricks are what sets these barns apart, allowing for ventilation and light within the barn. The brick walls had to be carefully constructed. The exuberance of the design had to be tempered by any miscalculation in the arrangement and number of openings so as not to weaken the barn's structure.

The multiple projecting wings of this barn make it an attractive sight along Route 57 in Mansfield Township, Warren County, New Jersey.

COMPLEX BARNS

Driving through Warren County, New Jersey, one sees Pennsylvania barns forming the core of beautiful farm complexes. The term is shorthand for two-level forebay bank barns common on both sides of the Delaware River and is not intended to slight New Jersey barn enthusiasts.

The two views on the left are of the same structure from different angles. This barn complex along the road outside of Rockport, Warren County, has a stone bank barn with round ventilators and a round date stone niche in its peak and a frame barn surmounted by a large cupola.

Sometimes it's just how the light hits a barn or the combination of colors that makes us admire a barn as in the State College area of Centre County.

BARNS AS CHARACTER DEFINING ELEMENTS OF THE LANDSCAPE

This barn, built into a hillside in Washington Township, Berks County, has a posted, extended forebay on the left and a bank shed on the right.

A dark silhouette against the sky at dusk, a humble structure on the corner of an original homestead, a cluster of buildings on a forgotten farm—the barn is a symbol of our heritage. It is a reminder of pioneers who built a future out of trees and soil.

How such large, often bright red, barns seem to gracefully fit into the landscape is a mystery. Perhaps it is more accurate to say that they have become integral parts of the landscape. They can project a sense of timelessness and a farmer's connection to the land that supersedes their aesthetic qualities. They help define the rural landscape. Farmland preservation should not ignore historic farm buildings. Preserving the barns in the fields and the fields around the barns should be done in concert.

Sometimes it's the setting that makes a barn so appealing. One cannot help but notice this farm vista while traveling along the Pennsylvania Turnpike in Lebanon County.

The oft-photographed Burgess Lea barn, Solebury Township, Bucks County, is one of a handful of double barns found in the county. The stable doors on the older, left hand, section have keystones above them matching those on the house.

While more typical of south-central Pennsylvania, barns with double stone outsheds are found in portions of Berks County as in this barn near Rancks in Caernarvon Township. The three sets of barn doors suggest a five-bay barn containing three barn floors or threshing bays.

The late afternoon sun adds color to a barn in Virginville, Berks County, with an unusual large opening in the gable.

Stubborn and still standing: The countryside is filled with many old, dilapidated barns. Like many others, I'm always happy that it's still standing —stubborn and still standing against all odds. As long as they're still standing…

Opposite: A photo of a barn in Solebury Township, Bucks County, from 2010 before it was restored (above). The barn has a 1796 date in the peak of the gable. Even as tired as this barn looked, its handsome jack arches and large keystones speak of its original grandeur. It seems fitting that the "after" picture evokes the reawakening of spring time.

This barn in Limerick Township, Montgomery County, has a nicely staged effect that makes it a traffic stopper!

Like many others, the paint on this Mail Pouch barn near Carlisle, Cumberland County, is slowly fading away.

Grain bins in formation make for an interesting sight on a wood frame barn near Mozart, Bucks County.

Shadows against a barn near State College, Centre County, make barns look alive.

Two views of a barn in Buckingham Township, Bucks County, painted in brilliant yellow standing out in a world of red or white barns.

By the twentieth century, the gambrel roof, which provided more loft space than a gable roof, became popular. The barn is enhanced by a wonderful stone silo in Lower Saucon, Northampton County. The reason this is one of my favorite barns is because blue sky, white clouds, and red tractor combine to make this one of my favorite photographs. It is also a good example of a gambrel roof barn and an unusual stone silo.

Gambrel roof barns became extremely popular, and pre-cut kit barns could be ordered by mail from Sears and other companies, such as this barn in Hunterdon County, New Jersey.

DETAILS

ARE WHAT MAKES A BARN SPECIAL

The 1810 Adam & Catharine Cressman barn, West Rockhill Township, shows the transition from splayed loophole vents to wooden windows in the early nineteenth century. The wooden date board simply adds more character to this handsome structure. This barn is one of the few barns where the bank is on the south side of the barn. The forebay was added a later date. The barn was constructed as a closed forebay standard barn. Adam Cressman was a carpenter living in Rockhill when he purchased the farm in 1799 from his father Abraham who lived in Upper Salford, Montgomery County. Born on January 23, 1774, Cressman was about 36 when the barn was built.

The handsome stone corner piers and the faded paint on the stable doors and forebay wall with its arches makes this barn in Bedminster, Bucks County, feel "real". It was a little spruced up for a barn tour.

Many barns in the Berks and Lehigh County region have decorative painted bands along the bottom of frame superstructures.

Barn Stars and Flowers

Many Pennsylvania German barns have large geometric designs painted on them. Most are six, eight, or twelve-pointed stars painted within a circle directly onto the barn siding. There is no evidence that would suggest that these symbols were used for anything more than to beautify a barn. In the early twentieth century, Wallace Nutting wrote that these decorations had a superstitious meaning, and they became known popularly as hex signs, a term which is now out of favor among scholars.

Here are two views of a brick barn in Upper Hanover, Montgomery County. The brick barns with open brick ventilators are much more common in the center of the state than in the southeast. A few examples of this exist in northern Montgomery County.

This barn, dated 1849, in Lower Saucon, represents one of the finest examples of a Pennsylvania German barn in Northampton County.

The barn has winnowing doors on the forebay side, opposite large wagon doors on the bank side. These doors were originally designed to create a draft when threshing of grain was done by hand. This barn has two threshing floors, each with its own set of winnowing doors. The barn stars and white arches make this barn a true work of folk art.

Many Pennsylvania German barns are adorned with barn stars or arches. The owner of this barn in Salford Township, Montgomery County, went a step further when he painted a false window with an arch above it over a pair of winnowing doors to provide a little more balance to the barn's facade.

This brick barn in Colebrookdale, Berks County, with its cracked wall, rusted metal roof, and faded barn stars, still evokes its proud heritage.

Here are a series of views of a barn in Bedminster with character as well as resident 'characters.' The fact that this is a closed forebay barn altered with an extended forebay and side addition seems less important than the fact that it contains numerous small features that tell the real story of a barn.

Some barns are special not for their architecture but the stories they quietly tell.

Same barn, different time, different animals posing...

A basketball hoop, industrial-style light, and peeling paint make this an eye-catching scene. Although this barn is located in Bucks County, Pennsylvania, I can imagine it from a scene in the movie "Hoosiers" where a young man practices his shot into the night.

A barn in Brecknock Township, Lancaster County, with a headless horse weather vane, swallow hole, barn star, and feed sign… what more can one ask for when barnstorming?
Opposite: An intricate swallow hole in Bucks County.

Swallow Holes, Martin Holes, Owl Holes, and Ventilators

If you look high up in the gable ends of many Pennsylvania barns, particularly those with a Pennsylvania German heritage, you will see holes cut between the planks. Many are simple diamond holes with smaller triangular holes at each point forming a Maltese cross. These openings are called swallow holes or *Schwalme Lecher* (the singular of this term is *Schwalme Loch*) in the Pennsylvania Dutch language.[7] Barns beginning in the second half of the nineteenth century were more elaborate. They were functional, ornamental, and symbolic forms. The holes helped with ventilation and allowed the entry of owls or other birds into the barn.

Alfred Shoemaker noted in *The Pennsylvania Barn* that these cut outs – "flat harts, Maltese crosses, crescent moons, tulips, etc., [appear] up in the gable ends of barns found in all parts of Pennsylvania."[8]

Two versions of Diamond Cross cut outs in Lower Salford, Montgomery County, and Moore Township, Northampton County.

A correspondent "WWM" gave the following advice to young farmers in the *Farm Journal,* May 1908: "If there is no swallow hole in your barn be sure to make one. Swallows catch myriads of winged insects that are injurious to agriculture. They destroy the caterpillar moths and the moth that lays the egg for the apple worm."[9] In addition to their practical function, it was believed that small birds prevented lightning from striking the barn.

These cut outs are not limited to Pennsylvania. They followed Pennsylvania Germans as they moved into Ontario, Canada. One study of swallow holes, "The Diamond Cross: An Enigmatic Sign in the Rural Ontario Landscape," by Thomas McIlwraith looks at "the how and the why" of these cut outs.[10] It includes a number of observations that are applicable in Pennsylvania. It notes that diamond crosses are found on farm outbuildings only. They are symmetrically placed and always well above the eaves line. They occur singly, in side-by-side pairs, and less commonly in groups of three or more. Single crosses account for more than half of all examples.

The physical aspects, "the how," of the early Maltese crosses are easily explained. The diamond is most often cut from two vertical boards. They are cut out with corner angles of 45 or 90 degrees. A plain diamond is the easiest of symmetrical shapes to cut into planks… It takes an extra effort to extend the diagonals, punch out the side triangles and notch the other triangles' top and bottom. Diamond crosses are peculiarly suited to board siding, and the design is poorly suited to the masonry construction common to European farms, making Old World precursors hard to trace. He observed that sometime after 1880, the use of the diamond cross died out. Very few appear on barns built after 1894, and only one known to have been built after 1904.

The "why" of their construction is a more difficult question. They are generally called "swallow holes". The swallow is a harbinger of spring and when it flies low over water or around poles, a signal of impending storms. There was no evidence tying these abstract uses to the shapes of openings, so it is hard to reconcile shape and purpose. McIlwraith suggests that, "any utility of the diamond cross, beyond conveying a message, must be considered a consequence and not a cause of its existence."

If not to attract swallows, were they for ventilation? Again, ventilation does not seem to be their primary purpose. The diamond cross, and most certainly the later, more ornately shaped holes were not really very useful for light or ventilation. Barn ventilation either came from roof top ventilators or simply relying on the shrinkage of barn siding to open cracks through which air could readily circulate. Similarly, farmers could get more light while they were in their barn by leaving doors open.

So, we don't know if they were spiritual, functional, or decorative. Perhaps the lack of documentation serves as a reminder of what we no longer know about our rural heritage. They may just fit into the Pennsylvania Dutch concept of *yuscht fer schee* (just for nice), in which artworks can be appreciated for their pleasing aesthetics alone, without the necessity of a deeper meaning. Either way, I always look up to the peak of a wooden barn when out barnstorming.

Stone barns throughout southern Pennsylvania had a round hole, 4-6 inches in diameter, usually in the barn peak. They are generally referred to as owl holes, allowing easy access for barn owls who helped rid the barn of mice and other rodents.

Left to Right: An ornate tulip cutout in near Lititz, Lancaster County, and simple diamond in Plumstead Township, Bucks County. The small birds on the top of the barns may have an opinion as to whether or not these are swallow holes. Notwithstanding the above, these holes are different than pigeon holes (Dovecotes) which were often built into a wall to house pigeons either as a meat source or for pets.

Many farms encouraged pigeons to roost in their barns. Hinges were located at the various levels of the cote to gain access to the birds. Photo taken in 2015 before recent alterations to the barn.

The May 1914 edition of the *Farm Journal* had an article "Pigeons For Profit" as a response from an earlier article that appeared in the magazine.[11] Starting with three pairs of pigeons, a correspondent had 51 squabs in the first year. The magazine also had an advertisement for their "Squab Secrets" booklet. The booklet notes that pigeons were bred for pets or meat and the two should not be combined. Pigeons bred for business "from a pot-pie standpoint" were dressed, cooked, and served the same as quail. While common pigeons could be used, and sometimes did well, farmers were advised to improve the birds by mating them with Homer pigeons.

Light pours through the wooden gable boards and the swallow hole at the Sunrise Mill barn in Montgomery County.

DOORS, HOODS, & WINDOWS

It is a shame that barn builders did not explain their reasons for building the way they built. Hoods over doors could serve several purposes. They could be to protect people entering or leaving the building from snow, rain or sun. They could protect the wooden door from the same elements. They could also be purely decorative.

Hoods over doors on houses certainly make sense to protect people, such as visitors, from the elements. The farmer would get wet after taking one step away from the door, so that doesn't make too much sense. Most of the barns that have hoods are stone barns, so the idea that they would protect a wooden door from weathering makes some sense. I like the last option best. They looked good!

Above: This Dutch door, with its barn star and simple projecting hood, is located in the gable end of a barn in Haycock Township, Bucks County.

Opposite: Another interesting door hood in Upper Saucon Township, Lehigh County.

If you take the time and look carefully, the ghost of a barn star on the stable door hood of a barn in Tinicum Township, Bucks County, can still be seen.

Above: The gable end of this barn in Bedminster Township, Bucks County, has two different sized doors with scalloped-bottom hoods. A handsome paint scheme highlights the doors, hoods, the window above the wider door providing light into a granary, and the small, shuttered window.

Right: This barn in Solebury Township, Bucks County, has a larger and a narrower door in the gable end, but both doors have hoods with segmental bottoms.

Taking attention to detail one step farther, this barn in Lower Saucon, Northampton County, has door hoods with cyma reversa contours and barn stars for additional ornamentation.

The ghost of a hood showing how they were attached to the barn with projecting outlookers in Tinicum Township, Bucks County.

The owner of this barn in Greenwich Township, Berks County, apparently wanted symmetry and painted a false door on his wagon doors. This seems like asking for confusion when going to the barn early in the morning or late in the evening.

Red Paint and White Arches

Red paint is composed largely of ferric oxide, also known as rust, a cheap ingredient for paint mixing. This additive has certain anti-fungal qualities useful for preventing wood rot. Many thrifty farmers painted their barns this economical color and passed the tradition on to create the ubiquitous red barn seen throughout North America. Red barns often have doors and windows accented with white paint. One theory about white painted arches is that they were not just for looks: their light color in contrast to the red background enabled the farmer to see the barn doors in the dark![12]

Open and Shut: two views of the same barn doors in Solebury Township, Bucks County, showing the open service door, sometimes called a “man-door,” which in this case features a distinctive profile with curved “ram’s horn” hinges.

Above: This barn in Northampton Township, Bucks County, has a simple beauty. The granary door is white on a red background while the stable doors have the opposite color pattern.

Right: The irregular spacing of the battens due to the random width siding and the broken window gives personality to this barn in Washington Township, Northampton County.

Two barns in Berks County:

Above: The weather worn wagon doors with a door-within-a-door evokes years of service for this barn in Washington Township.

Below: This service door has a simple rounded hole for access to the door latch. The Earl Township farmer really wanted to secure his barn, so he added this large conspicuous locking system.

The simple coffin-shaped service door and rough siding add character to this barn in Richmond Township, Berks County.

Right: On the bank side of the barn are large wagon doors with smaller service doors. The doors on both barns have attractive rams horn hinges and the corners are clipped.

The barn has a large date stone in the east gable end that indicates that it was constructed in 1820. The grace of the arched stable doors and windows in the enclosed barnyard in Newlin Township makes this one of the most attractive barns in Chester County.

The Audubon barn in Montgomery County has an arched wagon bay entry on its side elevation.

Integral wagon bays are rare in southeastern Pennsylvania, particularly so with nicely arched openings. A walk around to the barnyard side of barn shows a simpler flat design cut into the frame forebay wall.

Barred openings were the norm for providing light into the stables of early barns. The glass window above was a later improvement in this barn in Tinicum Township, Bucks County.

An example of a barred window with an exterior wood shutter visible in Lower Saucon, Northampton County.

Above: This Berks County example is enhanced by a harness peg and tapered battens on the shutter.

Below: Wooden shutters or small doors would be closed when necessary. The small opening is under the forebay in the barn below, in Lower Saucon Township, Northampton County. The metal track above was for a manure carrier and the larger door is the top half of a stable Dutch door.

It is always a treat to see logs peeking out from what initially appears to be a plain wooden structure, such as this barn in Tinicum Township, Bucks County, also shown during various stages of renovation on the following pages.

STEPPING INSIDE AN OLD LOG BARN

Log barns are not extinct—but they are an endangered species. Log barns were common during the eighteenth century. According to a compilation of the 1798 Federal Direct Tax of 1,395 barns in upper Bucks County, 955 (68 percent) were constructed completely of log and a total of 1,072 barns (77 percent) were at least partly log.[13]

A broader analysis of log barns is difficult because the 1798 tax records do not exist for the southern half of Bucks County. One of the few central Bucks County municipalities for which 1798 data exists is in Warminster Township where 21 (47 percent of the total) were log barns, 17 were stone barns (39 percent), one barn was log and stone (2 percent), and 6 barns (13 percent) were frame barns.

Similarly, in Upper Makefield Township a 1796 tax showed that of the 79 barns there were 29 log barns (37 percent of the total) in 1796. The remaining barns were 35 frame barns (44 percent), 1 frame and stone barn (1 percent) and 12 stone barns (15 percent) while 2 barns (3 percent) had no material noted.

Another view of a log barn in Tinicum Township, Bucks County during a recent renovation. Barn siding has been re-applied to the structure. The empty joist pockets define the top of the stable from the hay mow above. The photograph above shows how this ground barn was used. The stable logs have chinking and daubing between the logs to provide protection to the animals within, while air was allowed to penetrate the upper mow to ventilate the hay.

Putting log barns into context of ethnic affiliation is also problematic. Montgomery County 1798 Tax Records for barns and outbuildings are limited to the southern half of that county. Alfred Shoemaker's *The Pennsylvania Barn* provides numbers of specific barns in portions of Montgomery County. None of the 14 municipalities had 50 percent of their barns described as log. Only two municipalities having between 42 percent and 46 percent log barns. Like Bucks County, the southern half of Montgomery had a higher percentage of non-Germans than the northern part of either township. An analysis was done by Alfred Shoemaker for his book, *The Pennsylvania Barn*, that suggests 28.79 percent of barns in 14 townships at that time were log.

Tax lists for Gwynedd Township label eight of the twenty log barns as "old". In Whitemarsh Township four of the fourteen log barns were noted as old. Many log barns

The same barn in Tinicum Township, Bucks County, featured on the opposite page with no exterior indication of its log structure under the siding and the many shed additions.

in other townships are also described as old and the highest percentage of "old" barns were log. This supports the commonly held idea that the first barns in communities were built of log and were replaced by other barns afterwards.

Log barns provided pioneers with adaptive advantages including:

- No metal nails were needed to secure the logs at the corners.
- The logs provided protection from harsh weather.
- Logs were readily available.
- The weight of the structure was carried by the corners.
- Basic tools carried by all pioneers were used in log construction.
- Initial structures could be built quickly.

Five views of a log barn in Bethel Township, Berks County. The logs are “discovered” when the doors are opened. This barn has dovetail notches. Most barns in Bucks and Northampton counties have steeple or V-notches as seen on the example in Tinicum Township, Bucks County, on the opening page of this chapter. One photo is taken from one crib looking across the threshing floor. You can see the opening in each log wall to pitch hay into the designated crib. Pegs were inserted between the logs to help keep them separated at the openings.

There are accounts of log barns in Philadelphia in the eighteenth century. The January 14, 1789, *Pennsylvania Gazette* advertised the "property of Simon Cornell, deceased, situated in Lower Dublin township, containing 197 acres of land and a stone dwelling-house, a log kitchen adjoining the dwelling-house, and... a log barn with convenient stabling and other out buildings."[14]

The typical layout of a log Pennsylvania bank barn consisted of a double-crib barn plan with an open, central passage connecting the two cribs. The logs were often hewn to be almost square, and the ends expertly notched to create a tight-fitting joint that locked the building together and looked good as well.

The common method of shaping logs for log barns requires hewing two vertical planes flat while the top and bottom edges are left "in the round." Hewn logs provide greater interior space than round logs.

Building a log structure with space between the logs, called chinking, is easier than building with no space between the logs. The top and bottom of the logs may be left round, and the notches do not have to be as precise, requiring greater carpentry skill and time. Most of the log barns have closed or boxed corners where the logs do not project beyond the planes of the building.

The most common corner joint is the peaked saddle notch or V-notch (occasionally called a steeple notch as the "V" is inverted). Half-dovetail notches are also found throughout Pennsylvania.

The portion of log barns at the ground level, sheltering animals, have the spaces between the logs filled with chinking, while the spaces between the second story logs are left open for ventilation.

Log barns had limitations. Log structures have finite limits to their size. The weight of the logs limits the height of structures. The length of logs limits the length of walls; splicing logs together is difficult and introduces weakness in the wall unless a medial post is used.

Also, logs naturally taper in circumference from one end to the other, hewn logs were arranged large end on small, and vice versa, to keep the walls level. Logs were stacked higher than they could be lifted by hand. They were raised by ingenuity of the builders without any complex machinery.

In the essay entitled "The Log Barn" in *The Pennsylvania Barn* edited by Alfred L. Shoemaker, Henry J. Kauffman asserts that the log barn was one of the most important architectural forms in Pennsylvania throughout the eighteenth and nineteenth centuries, but most of the early log barns have disappeared from the countryside.

Where did the log barns go? As farms developed, log barns were replaced by standard Pennsylvania barns. It is a treat when one finds a notched log from an original log barn repurposed for a floor joist. It has become one of the things veteran barnstormers check when looking at barns. Here are examples from Bucks County (above) and Perry County (below).

The graceful notching in this log barn in Washington County shows the beauty of these utilitarian structures.

Dovetail notches as seen on the barn in Bethel Township, Berks County, at the left are not as common as the inverted V-notches seen in the barn in Bedminster, Bucks County, on the right.

He suggests that the log barn was a "cheap and useful building; however, it was not an attractive one. It connotated pioneering with its poverty and hardships and thus a nation whose ideals were beauty and prosperity was quick to demolish the log barn and loathe to perpetuate this symbol of their early struggle." These structures were leveled to make way for "more attractive and functional buildings of stone or clapboard."[14]

Citing admittedly incomplete data, Kauffman notes that there were more barns built of logs in 1798 than all other materials combined. There were 6,813 log barns, 685 log and stone barns, and 43 log and frame barns out of a total of 15,835 barns. The number of log barns may actually have been much higher as there were 4,971 of unclassified construction material.

Today, as we move further from our early history, this symbol of the struggles of the pioneers has greater meaning. We have a greater appreciation of the hardships of earlier times, and of the skill required to build what only appears to be a simple structure. Our appreciation also extends to the trees of large diameter, whose long and straight timbers produced the sturdy log walls that have stood the test of time. As these barns disappear, they never can be replaced.

A view of the rehabilitation of a log over stone barn in West Rockhill Township, Bucks County, showing the stone stables on either side of a central wagon bay.

Log over Stone Barns

This particular type of barn is a variant of the ground barn, and is not included in either the works of Charles Dornbusch or Robert Ensminger who produced the standard classifications of Pennsylvania barns. Barn historian Alfred Shoemaker offers the term *Boddemscheier* as a one level or ground barn.[16] This term is often interpreted to mean "bottom barn," although the Pennsylvania Dutch term *Boddem* has several meanings including "ground, bottom, and floor."[17] Shoemaker doesn't use the more common term *Grundscheier* for a ground barn.

I have followed the renovation of a barn in West Rockhill Township for years after first seeing it in its original condition. Since that time, I have been in another one located in the same township and have looked for other potential examples.

The 1798 Federal Direct Tax for Hilltown Township, Bucks County, is one of the few sources that specifically notes barns as being "stone below log above" (ten barns) or "stone below frame above" (six barns). Many other township assessors will note barns as being "part log and part stone", "log and stone", or "stone and log" (including the Hilltown assessor), but there is no way to determine if the differentiation of the materials is horizontal or vertical.

In 1798, roughly 10 percent of the barns in the township were this type of structure and other recorded examples in adjoining townships suggests that this form was widely used, despite a dearth of documentation. These barns ranged in size from 20 feet by 23 feet to 50 feet by 38 feet.

A barn of similar appearance located in Tinicum Township, Bucks County.

Jowled posts, in the 1810 Cressman Barn in West Rockhill (top), and an 1818 barn in Springfield Township, Bucks County (opposite). *Courtesy of Brian Murphy.*

BARN FRAMING

TALL SHIP OR MEDIEVAL CATHEDRAL?

Part of what people feel about old barns is an admiration of the timber frame of which they were built. Even barns with stone walls had massive timbers. A hand-hewn timber frame inspires awe. We have to stop and appreciate the effort and skill needed construct and raise a large barn when all of the framing timbers were felled and squared up by hand. Even after sawmills made it possible to make square timbers by machine, all of the notching for the sophisticated joinery was still done by hand.

It is often the little things that can be overlooked, that make a barn special or that deserve our attention and admiration. One of the little features that make certain barns special are unusual posts and timbers.

Detail of the ornate treatment of a jowled purlin post on a barn shown on the previous page. The barn is dated 1818 in Springfield Township, Bucks County. ***Courtesy of Brian Murphy.***

A Jowled Post

For those who like to look at timber framing details, there is always a surge of excitement when we find a barn with jowled posts. A jowled post is a post that is wider at the top than the bottom. When most people think of posts, they presume that they should be wider at the bottom than the top, perhaps for balance or providing a sturdier base. Barn builders turned this idea upside down by putting the wider portion of the post at the top which provides a greater surface to support multiple timber members, most commonly both a wall plate and a tie beam at the same time.

Barn builders didn't begin the tapering of the timber at the end that eventually became the base and work their way to the opposite end. Generally, posts are relatively square until they approach the top. A short distance below the top end, the timber is flared wider either in a long taper or a shorter, more severe taper that resembles the stock of a rifle. This is the source of the term "gunstock post." There are a few examples where the transition is ornamented through architectural embellishment.

Swing Beam

One of the most interesting features of many barns is a heavy beam extending from wall post to wall post designed to create a clear span in the middle of the barn with no intermediate posts. They are typically located on one side of the threshing floor. In Pennsylvania, these are referred to as "swing beams" but are referred to as "bull beams" in other areas.

Swing beams were cambered (given a slight curve upwards) in the middle with the bottom edge straight and the top edge bellied up in order to minimize dead load deflection. There are two theories about the actual function of these beams. The word "swing" generally refers to the idea that a horse drawn hay wagon could turn or "swing" around on the inside without a post being in the way. If a smaller barn had standard bents flanking the threshing floor, the wagon could not swing around.

Another function is attributed to these beams. Creating a clear span in the middle of the barn allowed for the tethering of an animal in the center of the barn and walking them in a circle threshing out grain as an alternative to hand flailing grain.[18]

Above & Below: A swing beam can be seen in each section of a double barn, in Upper Makefield, Bucks County.

An arched swing beam in a barn in Buckingham Township, Bucks County, and a detail of the diminished shoulder providing extra support for a swing beam in Mansfield Township, Warren County, New Jersey. (Note the two pegs and marriage marks at the joint). This barn has a second swing beam which is partially visible under the closer one.

Above: A unique swing beam. Carved on the swing beam in the interior of the barn are the initials A D and the date 1846. Whether the A D refers to Amos Doan, the son of the owners, Benjamin and Sarah Doan, in 1817 or for the Latin term Anno Domini is unknown. One of the hidden features is the carved pinwheel on the underside of the swing beam.

Right: The same swing beam in Upper Makefield, Bucks County, boasts lamb's tongue detail on a chamfered swing beam. The bottom photo shows one of numerous pinwheels carved in the beams underside.

In smaller barns, as in this barn in Newtown Township, Bucks County, that measures 40 feet long, the interior space was maximized by this modification. The form of a central threshing bay or barn floor flanked by two similar sized hay mows was modified by having one large hay mow, and a smaller bay on the opposite side of the barn floor. The smaller bay was open at the bottom to allow additional floor space for increased storage of machinery or equipment and to allow hay wagons to swing around. The open plan was made possible by the installation of a large cambered swing beam supporting the floor joists of the extra upper loft space. This interior configuration can often be identified on the barn's exterior layout by the asymmetrical location of the central bay.

The barn in the second view below shows a more traditional three bay barn with two hay mows and a symmetrical elevation located in nearby Solebury Township, Bucks County.

Anchor Beams

The term "anchor beam" is generally used to describe the massive beams associated with the Holland Dutch barns of New York and New Jersey, forming the cross timber in the center "H" and supported on both ends by two posts. The most distinctive feature of anchor beams are the "through tenons" on each end. Pennsylvania does not have Dutch barns, but there are a number of barns with large, dramatic beams that evoke the feeling of anchor beams.

These are beams that generally support the scaffold or overden located in the central bay. They run parallel with the ridge and are often arched. Like true anchor beams, their tenons protrude through the end posts.

Vertical purlin posts are joined with a horizontal beam or collar in this barn in Nockamixon Township, Bucks County. In German terminology this unit of the framing is described as a "*Stuhl*" (chair) and the angles of the posts determine whether the frame is standing (*stehender Stuhl)* shown above or lying (*liegender Stuhl)* as seen in the photograph on the opposite page.

There are two traditional German terms used by some architectural historians to describe two variations of early roof framing techniques in Pennsylvania barns. The term for the most common of these two forms is the *stehender Stuhl,* roughly translating to "standing chair." This describes a roof support system with vertical "standing" posts rising from the tie beams to the purlins, connected by a horizontal collar beam, creating the appearance of a bench or "chair." Although this is a literal translation of *Stuhl*, the more specific term *Dachstuhl* describes the framework or truss upon which the roof rests in common German building terminology.

The more dramatic looking system is the *liegender Stuhl,* literally "lying" or "inclined chair." In this heavy truss system, pairs of large, raking timbers following

A view of the Bertolet Barn in the Oley Valley, Berks County, showing the early Germanic *liegender Stuhl* truss system working in tandem with a nineteenth-century innovation: the hay fork.

the slope of the roof extend from the outer walls to principal purlins, bearing the common rafters. These raking timbers are connected by a collar beam, and a straining beam, as well as a pair of diagonal struts. This later form is celebrated as a culturally significant feature, characteristic of early Germanic timber framing traditions brought to Pennsylvania.[19]

One advantage of this arrangement lies in creating an unobstructed space (without additional posts) above and below the lateral beams of the truss. It is likely that this system, dating from at least the sixteenth century in Alpine Europe, was developed to support the immense weight of tile or thatched roofing materials common in early Pennsylvania, and used throughout Europe to the present day.[20]

The inside of a wooden latch from a barn in Gettysburg, Adams County. Half round holes, used to insert your hand and unbolt a door from the inside, are much more commonly found in south central Pennsylvania than in the southeastern counties. Many of these openings have swinging wooden panels on the outside to cover the openings.

BEAUTY FOUND IN SIMPLE THINGS

This spring latch is inset into the jamb of a brick barn in Jackson Township, Lebanon County.

Below, left: This door hardware shows decades of adaptation. The thumb latch appears to be from a house and a modern hasp was added later. Right: A standard thumb latch with a worn finger hole and wrought iron keeper has its own simple beauty.

Above: It is becoming increasingly rare to find a barn stable where old wooden stalls, stake mangers and feed boxes remain. Many of these items were replaced in the early twentieth century with the health regulations for dairying operations. Stalls with worn wooden stake mangers and feed boxes complete with holes to tether animals evoke the earlier uses of barns.

Right: Harness pegs and niches are often found in the part of the stable devoted to horses rather than cows. They are overwhelmingly located in the end of the barn closest to the house. They may have also been used to safely hang lanterns.

A double *Schpriggel*, as seen in this barn in Worcester, Montgomery County, is a rare configuration. A single *Schpriggel* is a more typical arrangement commonly found in horse stables.

Which End is Which?

Horses and cows were stabled in the basement level of Pennsylvania barns. In the vast majority of cases, horses were placed in the end of the barn closest to the farm house and cows at the far end of the barn. This orientation would provide easier access to horses which were of greater need to the farmer at different times of the day than cows, which were on a milking schedule. Not knowing when one might need a horse to go into town, the horses were not always allowed out in the pasture, but stayed in their stable.

Pennsylvania barn historian Robert Ensminger noted that the horse stable was generally located on the side of the barn closest to the house, as horses were considered more valuable than cows. Farmers wanted to be able to reach the horses more quickly in the event of an emergency which required evacuating the barn.

A barnstormer can generally find physical evidence for which side of a barn was for horses. Such features include the presence of small niches set into the stone walls by the stable wall doors for articles used in the care of horses. When these niches were placed next to the door nearest the house, they may have served as niches to safely place lanterns when farmers had to enter dark barns. Another feature related to horses were pegs to hang tack and harnesses when not in use. Many times, the pegs are missing, but the beams with large peg holes remain.

An ingenious feature found on stable doors is the retractable wooden bolt or *Schpriggel.* These wooden cross bars could be drawn across the door opening to confine horses in the barn while providing maximum ventilation beyond what could be provided by two-part Dutch doors. These bars are found on barns built by English and Pennsylvania Dutch farmers.

In defining Pennsylvania Dutch terms for barn features, folklorist and barn historian Alfred L. Shoemaker wrote that successful Dutchmen who had large barns always incorporated a *Schpriggel* described as "a wooden, two-inch square bolt the full width of a barn stable door, used to prevent horses from leaving stable when doors are open in hot summer months."[21]

In her classic *Barns of Chester County, Pennsylvania,* Berenice M. Ball (perhaps incorrectly) refers to these bars as "cattle bars", although they are generally only found on stable doors associated with horses. She cites examples near Phoenixville, West Chester, Marshallton, and Mendenhall.[22]

Left: Scholar of the Pennsylvania barn, Robert F. Ensminger, showing the action of the pocket door bolt or *Schpriggel*, and its function in allowing ventilation while keeping livestock inside.

Right: A *Schpriggel* usually has a metal ring attached like this one in Bethel Township, Lebanon County, to make it easier to pull out of the wall socket.

The stable level of a barn in Warwick Township, Bucks County, with straight harness peg and one unusual curved one. The curry box at the right of the photograph is located just as you enter the barn from a door facing the house.

Above: Harness pegs and curry box niche can be seen beyond this arched stall partition in Newtown.

Below: There is a peaceful rhythm to the feed troughs in this barn in Lower Saucon, Northampton County. Each stall had a hole to tether a cow.

SIGNS OF WORKING HANDS

The connection to the people who built and worked in an old barn is what makes a barn more than stone and timber. Every beam cut from local timber of pre-Civil War barns was made by the hands of a carpenter wielding a broadax. Every bent was raised by a group of neighbors or professional barn builders, and every stone was hoisted and laid by the hands of hardworking masons.

In all the barns I have been to over the years, the quintessential sign of working hands was found in a log barn in Tinicum Township, Bucks County. In this barn, the daubing in the chinking still shows where hands pushed the material between the logs. Several other views of this barn are included in this book.

Our barns are full of marks. These include marks used for construction, graffiti, and apotropaic (protective) traditions.

Above: A close examination of the hewn surface of a log crib barn in near Snyders, Schuylkill County.

Opposite, above & below: A hand print left behind by a builder on the surface of the clay daubing of a log crib barn in Tinicum Township, Bucks County.

Most eighteenth and nineteenth century barns have "carpenters' marks" on their frames which are articulated in Roman numerals. These marks were created with a race knife, chisel, or gouge, and used as tags to distinguish various elements of a timber frame and how the entire ensemble should be pieced together. These marks are formed using straight lines cut with a chisel or with the u-shaped, or scooped, end of the race knife. They are also called "marriage marks" because of their pairing.

Layout lines and marriage marks not only help us understand how barn frames were erected, but also show the hand of the builder of the structure, which provides an added connection to the barn. Chisel marks formed numbers to identify the correct correlating timbers to use. Slashes or flags were often added to identify the appropriate bent for each timber.

The photograph below shows the layout lines and the marriage marks on a post in a Tinicum Township barn. It is a simple reminder of the artisanship that went into building barns.

Flagged marriage marks in West Rockhill, Bucks County, and Lower Macungie, Lehigh County. The markings in the bottom example show that this joint was the third joint from the end in the third bent while the one above was the second joint in the second bent. The top photo shows you can put a square peg in a round hole.

Above: With the widespread use of concrete in barns for cleanliness and health, barn owners and their families had a new medium to mark their presence. Hand prints, in addition to names, initials, and dates, became quite common, as seen in this barn in Buckingham, Bucks County.

Below: A precise name and date, clearly incised in the plaster of a granary wall in Solebury Township, Bucks County, by "Samuel G. Fluck, July 27th, 1853."

This Plumstead farmer wanted his cat to have easy access to the stable.

When I think of barns, I think of livestock, but I also think of cats, rats, and bats that call barns "home."

It is not just the signs of working hands that make barns special. Mice and other animals add their own modifications as well. Both barns (above & opposite right) are in Upper Makefield, Bucks County.

Right: A ground hog has moved into this underutilized barn in Lower Nazareth Township, Northampton County. He seemed mildly curious that someone was taking pictures of his home. While he looks cute peeking out of the barn, a groundhog can do tremendous damage to the underside of a barn by loosening the ground under the foundation on which the structure is built. A groundhog can move over 500 pounds of dirt in creating a burrow.

One of many bits of graffiti from the Stapleton Barn in the Oley Valley of Berks County, including the inscription of the truncated name "J x STAPLT" within a scribed circle.

Opposite: Mr. Koch wanted to point out his name to anyone who entered this barn in Upper Macungie, Lehigh County.

BARN GRAFFITI

Scribbled, scratched, or painted, barn graffiti ranges from simple written words to elaborate wall artistic renderings. People often left their traces in wet plaster. This type of graffiti often commemorates the construction or renovation to a barn or simply records a person's presence at a particular moment. Often this type of graffiti is dated and is left untouched for decades, offering a look into local history. Examples can be found ranging from the times of early settlement to the present day. Today, graffiti is often the last link to the people who worked long hours in barns; but it also can be understood as an expressive art form.[23]

Generally, as an illicit activity, graffiti in barns appears to show that a significant proportion of graffiti must have had some form of approval—even if only tacit. This, coupled with the elaborate nature of some graffiti, would seem to place it more in the category of folk art rather than clandestine scribbling. This has led to a recent increase in ways of trying to understand and interpret graffiti in a wider setting and suggestions of how to re-classify what is regarded as graffiti and what is regarded as art.

What forms do graffiti take? The following forms of graffiti may be encountered while barnstorming:

1. Incised – for example cut with a knife or similar blade
2. Scratched – using a point such as an iron nail
3. Painted – using pigment e.g., paint, ink, fabric dyes
4. Written/drawn – ink, pencil, charcoal, red chalk
5. Carved – using appropriate chisel into stone or wood
6. Punched – using a punch or similar pointed tool and hammered
7. Gouged – using a tool as a chisel, V racer or scorper

Graffiti found in the granary of a barn in Buckingham Township, Bucks County, including a dated inscription, with geometric figures and a house with two chimneys.

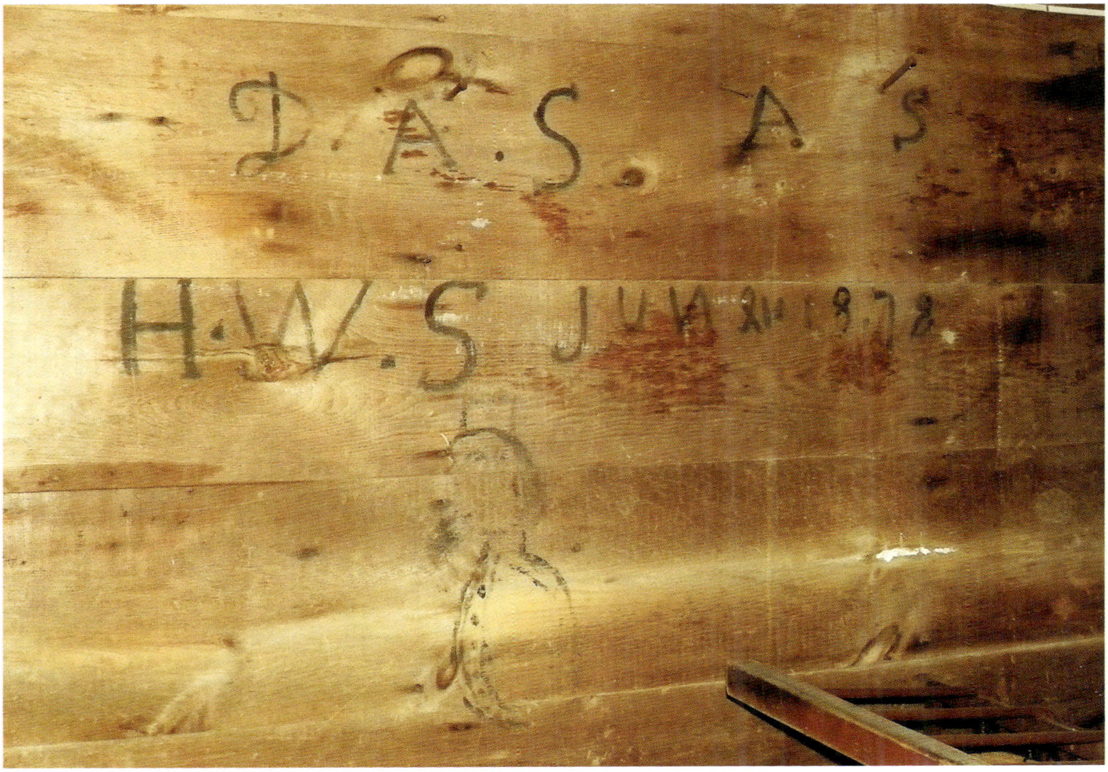

Above & Below: Someone drew a profile of what might be a soldier in this Newtown Township barn… and someone punched what looks like a snowman in this barn near Sellersville, Bucks County.

Two artistic renderings found in the stable of a barn in Perry County.

It is not unusual to find a series of names, initials, and dates all on the boards as seen on this threshing bay in Chester County.

While only a small part of a large structure, graffiti is fragile, vulnerable, and can be destroyed by a single inappropriate or unsympathetic treatment. In some cases, graffiti in its original location may provide additional context to a barn by directly informing us about building phasing or function. Its loss, due to its age and rarity or aesthetic merits, is always sad.

Another "community bulletin board" on the mowstead wall of a barn in Bucks County.

A large percentage of barns have incised marks. They are most commonly found in the mowstead walls. The six-petal rosette, also known as a daisy wheel, pinwheel, or hexafoil, is one of most common motifs. There is often a debate whether they have any apotropaic, cosmological, or religious significance, or if they played a role in the development of a building's construction, or if they are just a design easily made with simple tools. The pattern figure can be drawn with a compass, by creating seven interlinking circles of the same diameter touching the previous circle's center.

Above: Some of a number of rosettes and round designs in Upper Macungie, Lehigh County. Below: Rosettes and initials in a barn in Warwick, Chester County.

A political statement found in a barn in Bucks County

What caused a farmer to put his feelings about Grover Cleveland in his barn? What interrupted him from finishing his comment? One might presume that the writer was advocating Cleveland for a second term as president (Cleveland was the only president in American history to serve two non-consecutive terms in office, 1885–1889 and 1893–1897).

A farmer from Marion Township, Berks County, looks like he was designing a new and improved cream separator.

Above: Tally marks on a granary show that this farmer was storing buckwheat.

Right: Everyone wants to have the oldest barn in the area. Its hard to top this one!

Rittenhouse Barn, Worcester Township, Montgomery County.

A Message from the Past

There's a peculiar feeling that barn folks get when they wipe away layers of dust and cobwebs and see an etching from another era. Something as simple as initials and a date left to posterity by a resident of the farm show they built things to last, a sign of confidence in the future.

In a 2007 article in the *Reading Eagle,* written in advance of the National Barn Alliance and the Historic Barn and Farm Foundation's 2008 joint conference held in Kutztown, Berks County, Jim Lewars, administrator of the Daniel Boone Homestead, was quoted as saying that these inscriptions stand as monuments to the permanence with which the region's settlers approached life.

One can easily describe these inscriptions as a kind of message from the past. Who were these individuals? What was on their minds when they chiseled their names into the barn? Could they have realized their marks would have survived to the twenty-first century?

The questions may be unanswerable, but the survival of these inscriptions provides a link to those who came before us. "It's amazing, you can put your hand right where this person carved his name two centuries ago," Lewars said. "It's a real personal link to the people."

America's entry into the First World War undoubtedly influenced JM's creation of a flag in this barn in Tinicum Township, Bucks County.

Patriotic Graffiti

In Tinicum Township, Bucks County, there is a barn that reflects a young man's patriotism as the country was in the midst of the Civil War.

Courtesy of Bucks County MLS.

Near the southerly door on the east gable end of the barn is a very artfully carved inscription that reads "A. E. Worman July 4, 1861". On the bankside to the east of the barn door is an interesting political note. The carving reads "Union 35 States" carved with the same degree of skill. Strictly speaking there were only 34 states that remained loyal to the Union. Despite separating from Virginia in 1861, the northwest counties of Virginia that stayed loyal to the Union were not officially admitted to the Union as West Virginia until 1863.

Above: A carving of a ship, accompanied by the name "Samuel W. Paxson" on the stone. He was the son of the owner who was born August 25, 1863, making him around 14 when the stone was carved. Truth be told, this clipper ship is not actually on a barn, but a house next to the barn on pages 12-13.

Below: J L carved a Schweitzer-like barn with an 1836 date in Lancaster County and a subsequent owner used a horseshoe and a lot of nails to secure his rope.

The other type of tally marks often found in barns and houses are board-footage tally marks, created with race-knives by sawyers, indicating the quantity of footage in each board. The mark above indicates that there are 26 board feet, based on a system of shorthand used by sawyers in local water-powered saw mills. These are generally larger and less well cut than the Roman numerals used to aid in the joinery.

A working diagram: The carpenters working on the Hartzel-Strassburger Farm in Hilltown, Bucks County, sketched out the height of the bridge house (right) on the inside wall of the barn (above).

Most interior bents have internal ladders. This example in Lower Macungie Township, Lehigh County, is slightly unusual in that the ladder extends above the tie beam all the way to provide access to the roof. The rungs of the ladder are square on each end where they are set into the posts to prevent rolling of the rung while climbing. The cut out ventilator in the gable can be seen in the background.

THRESHING FLOORS & GRANARIES

A traditional barn had a three-bay plan—a central threshing floor or wagon bay flanked by hay mows. Hay was originally stacked loose and so the lower portion of the bents between the central bay and hay mows had a wooden "mowstead" wall to contain the hay within the mows. With the advent of baled hay, and then round bales, the walls were no longer necessary, and many of them are removed without a second thought. The mowstead walls are where the vast majority of graffiti was done. Unfortunately, much of the personal history of barns was lost with the walls.

A barn full of hay was the result of a lot of hard work and, in return, resulted in tons of satisfaction. Hay mows were open to the peak of the roof to allow for the storage of large amounts of hay. The central bays of barns often had beams laid on top of the tie beams of the adjacent bents to form an "overden" or "overbay". This upper hayloft above the center floor was known by New England farmers as the "scaffold." Nearly all barns (92 percent) contain a scaffold over the drive floor to store grain or late cuttings of hay. Some have a secondary scaffold at a different level.[24]

Above: This barn in Lower Saucon, Northampton County, has planks on top of the overden beams rather than the more common logs or beams.

Below: A small detail found on many overden beams are vertical pegs that prevent the logs from overly shifting, such as this example in Lower Salford, Montgomery County.

Overden is a term adapted from the Pennsylvania Dutch language term *Owwerdenn*, which is a loft over the threshing floor, from *owwer* (above) + *Denn* (threshing floor), sometimes also called an *Oberdenn*.[25]

Generally, the beams of the overden were not attached to the tie beams so they could be slid out of the way when not needed. The bottom picture shows vertical pegs that were occasionally used to keep the beams in place.

The grain produced by threshing was then stored a room built in one of the side bays or perhaps in a small loft above the threshing floor. Pennsylvania German barns generally had a room in the forebay for this purpose. This room, called the granary typically had a series of bins along one wall in order to store different grains. The boards could be slid out as the grain levels were depleted. The bins were on the inside wall and the room is usually illuminated by means of a small window on the frame forebay wall or outside gable wall. A number of barns featured doors to unload grain into wagons. Since these doors lead into the upper level of a barn they appear to open dangerously above the ground from the outside.

Granaries often had windows to provide light in the enclosed space.

Above: Keeping rodents out of a granary was a ceaseless task.

Below: While license plates were generally a preferred patch, this farmer had a supply of spiced luncheon meat cans he used to protect his granary in Lower Saucon, Northampton County.

Crossing the Threshold

"I should have known that" is the most common reaction one gets when giving barn tours and you point out the slats at the bottom of doorways that were designed to hold threshed grain from scattering beyond the threshing floor. Threshold is a "barn term" that has become a standard word in our language.

While we are familiar with the term when used in a household doorway, most people are surprised to learn the term's origin as a removable board about 6 inches high and extending form door post to door post and setting on the floor. When winnowing grain of its chaff, the threshold would "hold" the kernels of grain in the barn so the grain would not escape the threshing floor. Generally, they are found on the wagon doors, but are also found on winnowing doors and doors into the mows.

Most people are familiar with the expression "separate the wheat from the chaff" but don't actually understand its derivation. They may recognize the wheat portion of this expression, but most people are not as familiar with the word chaff. Chaff is the husks, or other seed coverings separated from the edible seed in threshing or processing.

Below: Threshold cleats in Upper Makefield, Bucks County (left), and another example (right) in Lower Saucon, Northampton County, with a pronounced transition to the threshing floor.

Hay Mows

Certainly, one of the main purposes of a barn was for the storage of hay. Hay was the fuel that ran the original "horse power" of a farm. When most barns were built, hay was stacked loosely in the mows on either side of the central threshing floor. Technology progressed to having hay pressed into rectangular bales and, more recently, into large round bales. What is often forgotten is that these bales are much heavier than loose hay and can put a strain on centuries old wood barn frames.

As the hay was traditionally stacked loose, rather than baled, hay-drop chutes provided a means to drop hay down to animals in the stables below the hay mow. Generally, the poles were simple smooth sections of trees. Some say this facilitated the hay dropping down. Others suggest alternately that timber is strongest in the round and also the quickest and cheapest, or that hay drops were built as an afterthought using whatever materials were available. Still others ask—why work the timber smooth when nature grows it smooth for you?

The hay chute in the Cressman Barn, located in West Rockhill Township, Bucks County, adds character to the barn's interior.

Two versions of the same shot, looking up a hay chute in Quakertown, Bucks County.

In order to reach hay stacked high in mows, ladders were often incorporated in the bents on either side of the threshing floor. To keep the hay from spilling into the threshing floor mows had solid mowstead walls. Foot holes were cut into the walls to reach the ladders.

There are a couple of neat things to see in the Stapleton Barn in the Oley Valley, Berks County, besides the half round foot holes. I like the sliding bolt and the round hole to reach the inside of the bolt in the doorway.

Common rafters resting on a continuous purlin in Perry County.
Opposite: Principal rafters and offset pricipal purlins in Lower Saucon.

Roof Rafters

Since many barns were very large, supporting the roof was always an important issue. The rafters that rested on the wall plates leaned in to meet at the roof ridge. Until the mid-nineteenth century, in our region, the rafters did not connect to a ridge pole or board but were joined together with an open mortise and tenon joint, sometime referred to as a "fork and tongue" connection. Roof rafters were generally sawn or sometimes left almost in the round. Barns either had common rafters, where all were the same size or they had a system with large principal rafters generally located above the interior bents.

Many barn builders provided intermediate support to the long rafters by installing purlin beams midway up the roof slope upon which the rafters could either rest or be tied into. Larger barns may have multiple purlins on each roof slope. The purlins could be continuous or interrupted.

This barn in Hilltown Township, Bucks County, apparently once had a windmill extending out from the roof ridge. Opposite, above: A barn in Doylestown has a cistern above the threshing floor. Below: This barn in Upper Gwynedd Township, Montgomery County, has a wooden cistern located on a scaffold at the peak of a stone gable wall.

WHAT IS THAT?

Making work easier—Windlasses

Pulleys made lifting heavy objects easier. One basic form of a pulley was a windlass which was a wooden cylinder with holes bored into it in an offset pattern into which handles were inserted and then removed to turn the device. Most people are familiar with windlasses with end cranks often seen on wishing wells.

The windlass in the winnowing door above made it easier to hoist items into the upper floor of a barn in Bethel Township, Berks County. The same barn has two windlasses in the carriage house (opposite, bottom) to raise wagon beds. Wheels were replaced with runners for use in snowy weather.

Above: Two windlasses to hoist heavy objects under a forebay in Tinicum Township, Bucks County.
Below: A pair of windlasses for servicing wagons in Bethel Township, Berks County.

Wicket holes

You will occasionally see barns with beams that have two holes located approximately four inches apart. It took me awhile to come up with the answer to what these mystery holes were for. Many barns seem to have them, often in multiple locations. These holes are vestiges of the location of bent hickory wickets that were part of the process of floating logs down the Delaware River to the mills in Bucks County. Hickory wickets were driven into individual logs to secure smaller logs that were laid perpendicular to the main log, thus, securing individual logs into a single raft. You are more likely to find these in barns in riverfront communities on both sides of the river.

If you look carefully above the sun pouring in through the ornate swallow hole you will see the metal rail of a hay track under the ridge.

Hay Tracks

One key feature for changes in barn size and design was the hay fork, or grapple, and introduction of the hay track. A hay track was a system where first a wooden track and then a metal track was installed along the underside of the roof ridge. Loose hay from the hay wagon was hoisted with a fork or grapple and usually moved anywhere along the longitudinal axis of the barn and dropped into a hay mow.

Prior to this innovation, loaded hay wagons were driven into a central bay of the barn, where it was forked by hand into the mows. In 1867, William Louden patented the "Louden Hay Carrier," a mechanical system used to load hay into the mows. In later barns, wagons didn't enter the barn at all. Barns had large doors in the gable or gambrel end and projecting hay hoods to accommodate the equipment.

An interior view of a vitrified tile silo, looking upward from below.
Opposite: Triple silos can be seen in Whitpain Township, Montgomery County.

SILOS

Did you ever peek into a silo and look up? I do it all the time.

"Silos can be constructed horizontally in pits, or vertically. Most silos of the first half of the twentieth century were vertical. Early silos were sometimes placed inside the barn, rectangular in shape, and of wood construction. These were quickly supplanted by round vertical silos located outside the barn, usually in a spot that would permit efficient filling (usually, from holes in the top) and unloading (usually, from a tier of doors from which silage was thrown down an exterior chute, which contained a ladder for access to the doors)."[26]

In the second decade of the twentieth century an article in the *Pennsylvania Farmer* magazine from May 1916 states "the silo has long since passed the experimental stage and has established itself as a necessary part of the equipment of the dairy farm."[27]

Above: This is the barn that gives Twin Silo Road in Plumstead, Bucks County, its name.
Opposite: A concrete stave silo in Upper Providence, Montgomery County.

In February of 1916, another *Pennsylvania Farmer* article recounts a recent survey of silos in Marion, Iowa, which found over 200 silos, of which 118 were "wood stave, 29 solid concrete, 27 flooring stave, 6 vitrified tile, 6 concrete stave, 4 pit, 2 brick, 2 concrete block, 1 modified gurler, and 1 stone. The stone silo was the first one built. It was erected in 1893 and is still good condition, but, aside from this one, the more permanent silos were largely built in 1914… Of the 94 silos built in 1914, prior to August 1 all were of solid concrete and 75 percent of all silo owners interviewed stated that if they were building again their choice would be solid concrete….[but] The vitrified tile silo is worth careful consideration wherever the tile can be secured cheaply enough."[28] The first six month's editions contain advertisements from twelve different companies making wood, metal, concrete, brick, and tile silos.

THE MAGIC OF A BARN

The sights, sounds, and smells of a barn
Dusty and covered in cobwebs
Dust dancing in the filtered sunlight

The effect of filtered light imbues a barn with a quality that exceeded just a space for storing hay. Something special seems to have seeped into the timbers of old barns.

An old barn is full of sights and smells that fill the senses.

These structures are like cathedrals. Enter into one and the space takes you over.

But they are also marvels of craft and function.[29]

"Something happens to the quality of light in a barn… it becomes softer, richer; it takes on the warmth of the beams."

- Jim Doherty,
A Barn is More Than a Building. It is a Shrine to Our Agrarian Past[30]

Sometimes the light catches little things in a barn such as the strands of a spider web giving them a simple beauty.

Something special happens to the quality of light in a barn. When there is a lot of dust in the air the light becomes softer and richer. Other times the play between the light rays and the darkness of the barn's interior adds to aesthetics of the barn's architecture. Light shining through gaps in the walls of an old barn or light shining through a window onto the hay gives it a golden hue.

Above: Sunlight filters through the siding of a log barn in Tinicum Township, Bucks County.

Above, opposite: Light playing through the brick ventilators of a Franklin County barn.

Right & left, opposite: These views highlight how pinpoints of light emphasize elements within dark barn interiors.

The Hodge Barn near Boalsburg, Centre County, is virtually unique with regard to the detailing covering its window openings.

FANCY BARNS

Generally, barns are very practical structures. Their simplicity is often what endears them to so many people. However, there are many barns that are more "fancy" than their function require. Some can only be described as extravagant with cross gables, arches, ornate louvers, and woodwork to meet the fancy of the owner.

The crowning element of many fancy barns are cupolas. Cupolas are small hollow structures at the peak of a roof designed to vent heat out of the loft area. They also keep hay in the barn dry by continuously circulating hot air as it rises. Cupolas were, undoubtedly, primarily functional when introduced, but they became a feature for adding beauty and proclaiming the success of a farmer. They often provide the most distinctive facets of a building. Even simple barns were often adorned with ornate cupolas as a means for a barn owner to express their individuality. Many farmers took particular pride in their barns. One only has to look at the multitude of nineteenth-century county atlases that contain lithographs of fancy barns.

Above: This barn near Duncanville, Blair County, is one of the best examples of a barn builder incorporating gothic arches, a projecting center cross gable and ornate octagonal cupolas to make a dramatic statement. Although only one is visible in the photo, there are two rear wings giving the barn an overall U-shape. Complementary outbuildings enhance the barn's setting.

This fancy barn is located near Cashtown in Adams County. It is currently located right on a public road, where one can easily imagine a barnyard on both sides of the projecting front wing. The barn has an integral wagon bay to the right, where the forebay and the mow on the inside are raised to accommodate wagons on the lower level. This was done on one side only to preserve more interior space in the upper level, instead of making the entire stable level this height. The barn is actually "T" shaped with a central large cupola and two cupolas on the main roof as well as the two on the front extension.

The evening sun washes the fancy barn at Fort Hunter Park outside of Harrisburg in warm sunlight.

Cupolas adorn many barns in Central Pennsylvania, such as this one in Cumberland County.

This dramatic barn in Washington County prominently displays a date of 1893. It exhibits applied decorative wooden elements more commonly found in the western part of the state than east of Harrisburg.

The barn ramp does not abut the lower portion of the barn wall which helps prevent hydrostatic pressure against the interior wall. The wooden bridge, like covered bridges over streams, is protected by a wooden structure called a bridge house.

Bedford and Somerset counties are the epicenter for applied wooden barn stars. The example below is readily recognized by those who travel the Pennsylvania Turnpike.

The end opposite the one in the above photograph shows the applied barn star and also indicates that there was originally curvilinear wooden gable trim of which only a portion remains.

Two farms, in Somerset (above) and Fayette (below) counties, remind us to drink milk.

The Round Barn in Biglerville, Adams County, has been adaptively reused as a market.

An interior view of the central tile silo within the Round Barn in Biglerville.

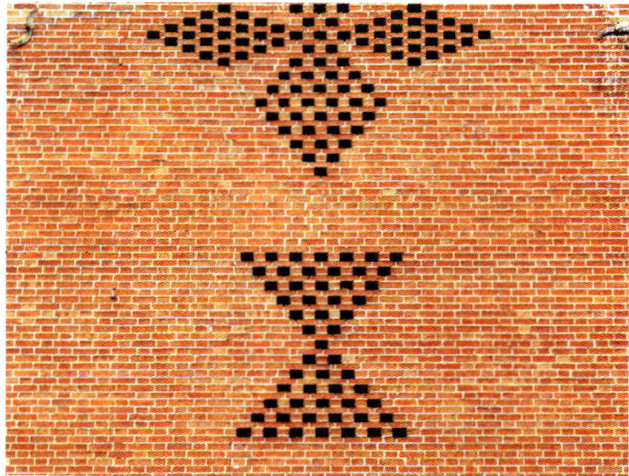

Above: This brick barn in Lurgan Township, Cumberland County, has a diamond shape pattern opening over the loading door of the rear outshed and a series of rectangular openings in the main mow. The star shaped anchor bolts above the loading door just adds to the beauty of the barn.

Right: Cumberland County is full of wonderful brick barns with patterned brick ventilators as with this example near Carlisle.

This Schweitzer barn in Franklin County sports sheaves of wheat and geometric patterns.

As Joseph W. Glass wrote in his book, *The Pennsylvania Culture Region: A View from the Barn,* in 1986 about brick-ended barns, they were "seldom painted, their natural brick color is highlighted by the shadowed contrasts of variously shaped openings in their brick gable ends that provide ventilation and light to the barn's interior". Their builders "faced the essential task of maintaining the structural integrity of their walls at the same time they were omitting bricks to form designs". These barns were "built to function well, to please the eye, and to last".

The decorative patterns produced by openings in the bricks consisted of Xs, diamonds, sheaves of wheat, Maltese crosses, goblets, initials, dates, and in one case a man riding a donkey.

RESPECT THE LANDOWNER
ASK PERMISSION BEFORE YOU HUNT
MEMBER
PENNSYLVANIA
WILLIAM R. HALE
139
W. R. HALE

BARN STORIES

Barns, like all buildings, represent people. They provide a connection to their owners and their builders. In addition to the marvel of their architecture, they tell stories.

You have to spend a few minutes when looking at the 1882 Hale Barn outside of Gettysburg, Adams County. Note there is a windlass (actually there is another out of sight to the right) built into the wall. Mr. Hale also incorporated a wheel rim for his hose reel (opposite). Old signs, new signs, a John Deere thermometer, as well as a cabinet with cherry tree spraying information, and a bulletin board all help tell the story of this barn.

This barn in Adams County appears simple but has a lot of features making it interesting to barnstormers. First, is the welcome sign! Walking up to a barn unannounced is always a bit nerve-racking. Add the four swallow holes in the gable, a cow weather vane, the wagon shed addition with a classic pickup truck and even the currently over-popular metal stars, and this barn is a treat to stumble upon along a country road.

An increasingly infrequent sign of rural life can still be found in Potter Township, Centre County.

Sometimes unannounced barnstormers are greeted with some skepticism.

Driving down a side road is often worth the extra time to get home. A lucky detour down a narrow road in the Mifflintown-McAllisterville area of Juniata County revealed this gem. This simple sign just seemed to speak volumes.

A farmer in Bedminster, Bucks County, shows his pride in being a member of the Pennsylvania Farmers Association.

A barn wall in Upper Providence, Montgomery County, includes a calendar from a farm insurance company, as well as a list of water depth calculations and repairs to the farm's well.

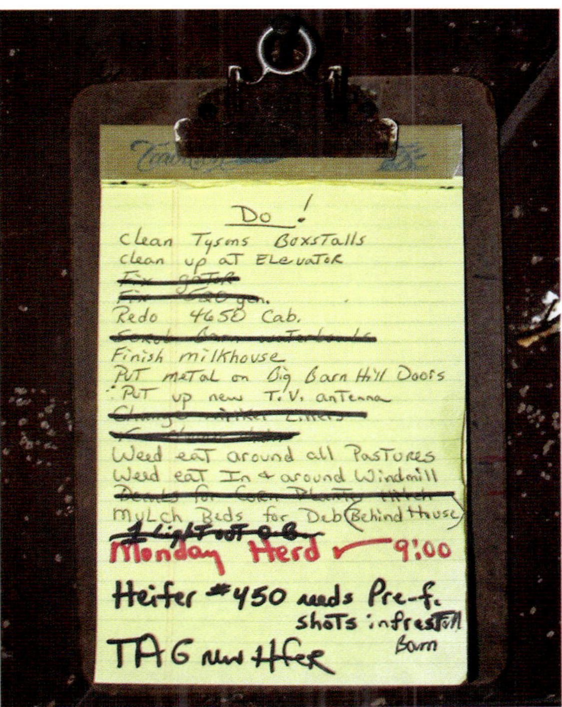

The farmer's "to do" list personalizes a barn. Daily chores, maintenance, building improvements, and livestock veterinary appointments are just some of the stories embedded in the setting of this Pennsylvania barn.

Nostalgia in Stone

Many barns have several date stones with different years prominently displayed. These are not graffiti where someone carved their initials and a date sometime after the barn was built but appear to be designed to commemorate the construction of the barn.

The gable end of a barn in Upper Makefield, Bucks County, has the inscription of B D 1839 three or four courses above two stones marked E D 1786 and A D 1786. Either it took forty years to finish the barn or Benjamin Doan wanted to commemorate an earlier structure that his family constructed by repurposing the stones.

This barn in Plumstead Township Bucks County, is built in two sections. The 1758 date may have been on the gable end of the original barn and was moved when the 1822 section was added.

What is a barn without a few tools hanging around? These too, tell the story of the farm and barn.

Tow chains and block-and-tackle assemblies suggest stories of long hours, heavy loads, wagon repairs, and perhaps even some maintenance projects for this barn in Lebanon County.

A patented 1921-1922 automatic watering dish for livestock suggests a broad range of updates and improvements to the barn based on agricultural regulations for dairy operations introduced in the early twentieth century.

The Elijah Funk Barn was the third largest barn in what is now Charlestown and Schuylkill townships, Chester County, measuring 70 by 35 feet and was originally valued at $350. Physical evidence shows that the current forebay on the Funk barn is a later extension. The dimensions of the barn, as noted in 1799, correspond to the dimensions of the original structure. Today the barn is slightly deeper than it was in 1799. The original floor joists are clearly visible under the forebay. They do not extend out to the bottom of the current forebay.

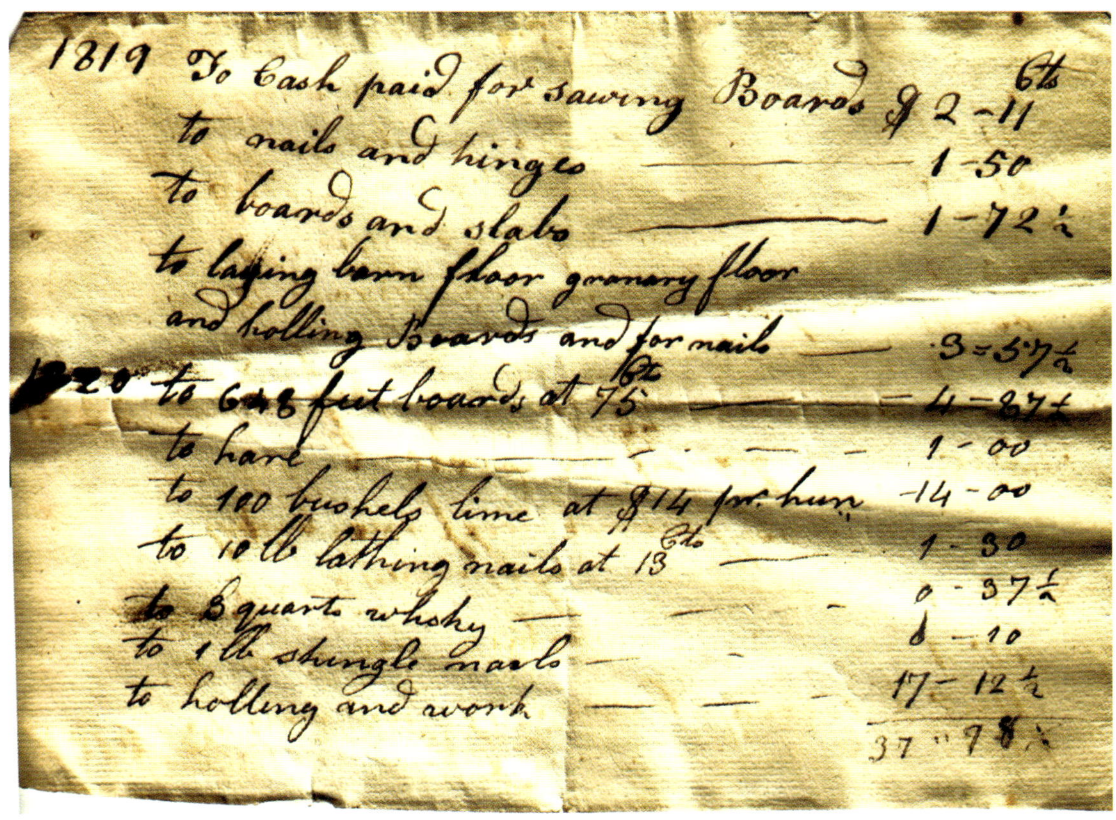

1819	To Cash paid for sawing Boards	$ 2-11
	to nails and hinges	1-50
	to boards and slabs	1-72½
	to laying barn floor granary floor and holling Boards and for nails	3=57½
1820	to 648 feet boards at 75 cts	4-87½
	to hare	1-00
	to 100 bushels lime at $14 pr. hun	14-00
	to 10 lb lathing nails at 13 cts	1-30
	to 3 quarts whisky	0-37½
	to 1 lb shingle nails	0-10
	to holling and work	17-12½
		37"78½

Building barns with "whisky"

It is rare to find original documentation for barn construction. This barn owner in Doylestown, Bucks County, had work done on his barn in 1819. The hand-written receipt contains entries for labor for sawing boards, laying the barn floor and granary floor, as well as for "holling" [hauling] materials. Purchased materials included nails and hinges, shingle nails, boards and slabs, as well as "hare" [hair], lime, and lathing nails for plastering. My favorite listing was the three quarts of "whisky"—a customary form of supplemental payment allotted for the enjoyment of the work crew.

Barnstormers can always find interesting details. Twentieth century dairying operations have their own particular items of interest. Here are a two views of a Jamesway brand manure track and trolley in the stable level of a barn in Buckingham Township, Bucks County.

Above: Here, you can see an open loading door on a barn in Adams county in the area between Gettysburg and Hanover. How many farmers used old bathtubs for water troughs?

Right: The McClean Barn in Gettysburg has a wooden spring latch that could be opened by reaching in through the half-round door opening. These types of openings are more commonly found in Adams County than other areas where metal latches were used.

"A man's barn bespoke his worth as a man. It expressed his earthly aspirations and symbolized the substance of his legacy to his children."
- **Berenice M. Ball, *Barns of Chester County, Pennsylvania***

This barn was located off the road in Upper Providence, Montgomery County. There was no one home, so I took this photo from the road. The barn was a nice example of a closed forebay standard barn but what caught my attention was the small sign above one of the gable entry door. "Faith, Family, Farming". I subsequently found out that this was a common saying and that such signs were common on the Internet. I also saw an article in which the president of the American Farm Bureau Federation stated it was the organization's "informal motto," but I had never seen it before.

Farmers learn early that you can't control Mother Nature. Farmers face floods, drought, insects, and disease. Farmers plant money in the spring and hope it grows. They persevere with hard work, the help of their family, and faith that that will be enough.

A farmer does not build a barn that is designed to last for centuries as a convenient structure for immediate farming needs. It is more than an economic decision. A farmer builds this type of barn with the hope and faith that the family will benefit from this expenditure well beyond the farmer's lifetime, that the land will be fertile and the family will be on the land for many generations.

Whether a dark silhouette against the sky at dusk or a simple structure seen along a road, the barn is a symbol of our heritage. It is a reminder of pioneers who built a future out of trees and soil. We will only save what we respect. Hopefully, learning a little about these barns will garner the deserved respect.

Happy Barnstorming!

Jeffrey L. Marshall

ABOUT THE AUTHOR

Jeffrey L. Marshall, president of Heritage Conservancy in Doylestown, Pennsylvania, has been preserving land and historic resources at the conservancy for 40 years. Heritage Conservancy has been a leader in protecting the character of the community as well as the natural resources that sustain it as important elements in creating livable communities. Jeff's efforts have been at the national, state, and local level.

Recently, Jeff orchestrated the acquisition of the former David Library of the American Revolution, in Washington's Crossing, and its subsequent transfer to the Delaware Canal State Park. As past president of Heritage Conservancy, he oversaw the organization's numerous land preservation activities. He personally prepared open space plans for four municipalities leading to the preservation of over 5,000 acres to date, beyond the Conservancy's own efforts.

Jeff began his career in historic preservation in 1976, and has been involved with the documenting, photographing, and researching of 10,000 old buildings. He is a recognized expert in southeastern Pennsylvania historic architecture. He is the author of books on Bucks County barns, Bucks County farmhouses and several township histories. He has taught courses on history and preservation at the Bucks County Community College. Jeff has been the featured speaker for numerous historic preservation programs and tours in Bucks, Montgomery, and Northampton counties, as well as statewide and national conferences.

In his career, Jeff has served as President of the Board of Directors of the National Barn Alliance, was a founder and President of the Board of Directors of the Historic Barn and Farm Foundation of Pennsylvania, and formerly served as president of the board of the Pennsylvania Land Trust Association, and on the board of Preservation PA. He is a member of a number of local boards and committees. Jeff has successfully nominated over 40 properties and historic districts to the National Register of Historic Places.

Jeff was selected for a Lifetime Achievement Award as the "2015 Ambassador of Bucks County". He was the recipient of the inaugural "Bucks County Preservation Legacy Award" by the Bucks County Commissioners.

ENDNOTES

1. Chisholme 2008.

2. Ensminger 2003: 109.

3. Bertland 1974, 18, 24-25.

4. Ensiminger 2003: 135.

5. "Historic Weathered Barns of Appalachia" *Appalachia Advocate.* <https://sites.google.com/site/appalachiachronicle/barns-bridges> Accessed 04-21-2021.

6. Beam, C. Richard. 2006. *The Comprehensive Pennsylvania German Dictionary.* IX:331. Center for Pennsylvania German Studies, Millersville University. Morgantown: Masthof Press.

7. Additional discussion of these classic forms can be found in a photo essay "Swallow Holes!" appearing in *The Forebay Post*, Spring 2019, Journal of the Historic Barn and Farm Foundation of Pennsylvania.

8. --- --- 1908. "Topics in Season" *Farm Journal.* XXXII (5):221. Philadelphia: Wilmer Atkinson Company.

9. Shoemaker 1955: 52.

10. McIlwraith, Thomas. 1981."The Diamond Cross: An Enigmatic Sign in the Rural Ontario Landscape" *Pioneer America.* XII:(1).

11. --- --- 1914. "Pigeons for Profit" *Farm Journal.* May XXXVI (5):326. Philadelphia: Wilmer Atkinson Company.

12. Literature associated with barn tours in Washtenaw County, Michigan. http://www.ewashtenaw.org/government/departments/community-and-economic-development/workforce-development/historic_preservation/Feb%2009%20site%20update/histweb/histweb_tours/barn_tours.pdf Accessed 04-21-2021.

13. Data from the 1798 Tax Assessment is taken from "The Barns of 1798" Shoemaker 1955: 91-96.

14. "To be Sold at public Vendue." *The Pennsylvania Gazette.* Philadelphia: Hall & Sellers. January 14, 1789 (3098):3.

15. Kauffman, Henry J. 1955. "The Log Barn" in *The Pennsylvania Barn,* Shoemaker 1955.

16. Shoemaker 1955. See also: Shuffleton, Frank. 1993. *A Mixed Race: Ethnicity in Early America.* New York: Oxford University Press. Shuffelton uses the term "*Buddem*" translated as "ground". *The Barns of the Midwest* (2018) by Noble & Cleek cites a work by Beaveau Borie stating that a ground level barn is known in the Pennsylvania Dutch vernacular as "*Grundscheier* or "*Boddem Scheier.*"

17. Lambert, Marcus Bachman. 1924. *A Dictionary of the Non-English Words of the Pennsylvania-German Dialect with an Appendix.* Lancaster, PA: The Pennsylvania German Society.

18. Online forums of barn enthusiasts include "Barn Geek" where the role of the swing beam is debated from several angles. http://www.barngeek.com/what-is-a-swing-barn-swing-beam.html Accessed 04-21-2021.

19. Ensminger 219-221.

20. Ensminger 219.

21. Shoemaker 1955: 46.

22. Ball 1974.

23. For an archaeological overview of historic graffiti, see Buglass, John. 2016. *Recording Historic Graffiti: Advice and Guidance.* Historic England. <https://pdfslide.net/documents/recording-historic-graffiti-advice-and-guidance-1-recording-historic-graffiti.html> For examples specific to Pennsylvania barns, see Donmoyer 2013: 77-84.

24. Mott, John. 1978. *Barns of New England: Surveyed by John A. Mott and Frank White for Old Sturbridge Village.* Unpublished Report.

25. "Oberdenn" is a term widely used by barn historians in Pennsylvania, mostly notably, Robert F. Ensminger through his work with the Historic Barn and Farm Foundation of Pennsylvania. The term is described in the HBFF glossary included in *Over the Blue Mountain: Log Barns of Schuylkill County.* Annual Conference & Tour Guide 2016.

26. Pennsylvania Agricultural History Project, http://www.phmc.state.pa.us/portal/communities/agriculture/field-guide/silo.html

27. --- ---. "Large Silo Equipment." 1916. *Pennsylvania Farmer.* May 16. XXXIX(19):11.

28. --- ---. 1916. "Silo Survey." *Pennsylvania Farmer.* February 12. XXXIX(7):13.

29. Many of these comments are quoted from a column: "A Broadside for Barn Preservation: Alternatives To Watching Old Barns Fall Down" by Charles Bultman, October 26, 2010.

30. Doherty, Jim. 1989. "A Barn is More Than a Building. It is a Shrine to Our Agrarian Past." *Smithsonian Magazine,* August.

A BARNSTORMER'S BASIC GLOSSARY

Anchor beam: Heavy beams tenoned through a pair of vertical posts, forming a distinctive "H" shape. True anchor beams are typical of Holland Dutch barns in New York and New Jersey, but Pennsylvania has numerous similar examples in barns built with an English cultural influence in Bucks County.

Bank barn: A two-level barn entered on the second level by means of an earthen ramp, sloped hill, or constructed bank against the barn.

Bay: An area or room inside a barn separated by framing units or walls, oriented perpendicular to the roof ridge.

Bent: A timber-frame unit of a barn constructed perpendicular to the ridge, which divides the barn interior into bays.

Double-barn: A barn constructed with two separate, often virtually independent sections, usually built at separate times, and sometimes built in different styles as distinct from a barn with multiple additions.

Double-decker barn: A barn with two levels above the stable level, usually consisting of an upper threshing bridge with sink mows on either side. The term is often confusing as these barns actually have three levels including the stable level.

Eaves: The bottom edges of a roof that overhang the wall of a building.

English barn: A single-level three-bay barn, most commonly of wood frame construction, generally 40 feet by 30 feet, with wagon entrances on each eaves-wall. This form is not common in Pennsylvania.

English Lake District barn: A two-level bank barn in Pennsylvania related to those found in Northeastern England. Usually includes a pent roof sheltering the front stable wall, rather than having the frame forebay characteristic of Pennsylvania barns. The exterior is generally built entirely of stone.

Forebay: A cantilevered or supported overhang sheltering the stable front wall of a Pennsylvania barn.

Gable: The triangular exterior upper wall of a structure, connecting the ridge to the eaves on either side.

Granary: An enclosed interior space within a barn used for storage of grain, usually containing built-in grain bins.

Ground barn: A single level barn, constructed of stone, log, or frame, usually with a central wagon bay, flanked by stables, and loft space above. Sometimes also called a "basement barn."

***Grundscheier*:** A Pennsylvania Dutch language word for a ground barn.

Hay drop chute: A light frame built within a mow to provide a means to drop hay down to animals in the stables below. Generally, the poles were simple smooth sections of trees which facilitated the dropping of loose rather than baled hay.

Hewn: Shaped from wood by use of hand tools such as a broadax.

Liegender Stuhl: German architectural term, translating to "lying" or "inclined chair." A characteristic feature of Germanic timber framing traditions in Pennsylvania, this heavy truss system employs pairs of large, raking timbers following the slope of the roof extending from the outer walls to principal purlins, bearing the common rafters. These raking timbers are connected by a collar beam, and a straining beam, as well as a pair of diagonal struts.

Loophole: A slit in the masonry of a barn, typically flared on the inside used for ventilation and light.

Marriage marks: Marks incised into the wood surface by carpenters at framing joints. Their function is to provide matching pairs for putting together framing members.

Mow: A bay where hay or straw was stacked, generally located on one or both sides of the threshing floor.

Pent roof: A single-sloped roof, anchored directly to the stone wall of a house or barn directly above the lower level access.

Pennsylvania barn: A two-level barn defined by the presence of two diagnostic features: a forebay or overhang sheltering the front stable wall; and an access to the upper level by means of a bank, earthen ramp, or bridge.

Pennsylvania Dutch (or Pennsylvania German): The descendants of German-speaking immigrants who settled in Pennsylvania between 1683 and shortly after the American Revolution, known for developing distinctive American traditions, architecture, agriculture, and a vernacular language (also called Pennsylvania Dutch). The word "Dutch" is historically of Anglo-Saxon derivation and was originally synonymous with German. The term was used broadly to include people of all regions where the Germanic family of languages are spoken, including Germany, Switzerland, Austria, Alsace, and the Netherlands. In Pennsylvania, the term "Holland Dutch" specifies those from the Netherlands.

Pennsylvania Dutch (or Pennsylvania German) barn: A cultural designation given to a barn produced within the Pennsylvania Dutch cultural region, not specific to type or form, although sometimes used to imply the classic forms of a Pennsylvania barn.

Purlin plate: A horizontal beam running parallel to the roof ridge in a timber frame building, used to support the rafters, generally at mid-span, of a roof structure.

Schpriggel: A Pennsylvania Dutch term for a wooden retractable door bolt, approximately three inches square, that is used to prevent horses from leaving the stable when doors are open in hot summer months.

Schweitzer barn: A Pennsylvania Dutch language term for the original form of the Pennsylvania barn descended from Swiss prototypes. In its classic form, the Schweitzer barn features a cantilevered frame forebay that produces a distinctive asymmetrical roofline. Transitional and extended variations of the Schweitzer barn may include outsheds or other additions that change or balance the appearance of the roofline.

Stable front wall: Stone or frame wall under the forebay of a Pennsylvania barn, featuring stable doors and typically designated as the front of the barn.

Standard barn: Form of the Pennsylvania barn characterized by a balanced bent configuration, and symmetrical roofline.

Stehender Stuhl: German architectural term, roughly translating to "standing chair," describing a roof support system with vertical "standing" posts rising from the tie beams to the purlin plates, connected by a horizontal collar beam, creating the appearance of a bench or "chair."

Swallow Hole: An opening cut in various geometric forms, typically on the upper gable, which admits birds into the upper level of the barn while providing increased ventilation and a decorative element to a barn. Geometric swallow holes are commonly associated with the Pennsylvania Dutch diaspora in the Mid-Atlantic, Midwest, and Ontatio, Canada. In New England, circular holes in the gables of barns are commonly called "martin holes." Swallows and martins are members of the avian family *Hirundinidae.*

Swing beam: A large interior transverse framing member located approximately 7 feet above the barn floor, spanning the entire width of the barn, supporting framing members above, with no framing below. Associated with the need for a turn-around for teams of horses in threshing, loading, or navigating the otherwise narrow area of the threshing floor. Many larger examples are cut to induce a camber (arched shape) in the center.

Rafter: One of a series of internal structural members forming the diagonal slope of the roof from peak to the eaves.

Threshing floor: Designated area within a barn where the hand-processing of grain took place, typically adjacent to the granary, and flanked on either end by doors used for ventilation.

Truss: An assembly of framing members forming a rigid support structure, such as that supporting the span of a roof, perpendicular to the ridge.

BIBLIOGRAPHY

Ball, Berenice M. 1974. *Barns of Chester County, Pennsylvania.* West Chester, PA: Chester County Day Committee of the Women's Auxiliary, Chester County Hospital.

Beam, C. Richard. 2006. *The Comprehensive Pennsylvania German Dictionary.* Center for Pennsylvania German Studies, Millersville University. Morgantown: Masthof Press.

Bertland, Dennis N. 1976. *Early Architecture of Warren County, New Jersey.* Belvidere, NJ: Warren County Board of Chosen Freeholders.

Borie, Beauveau. 1986. *Farming and Folk Society: Threshing Among the Pennsylvania Germans.* Ann Arbor, Mich: UMI Research Press.

Buglass, John. 2016. *Recording Historic Graffiti: Advice and Guidance.* Historic England. <https://pdfslide.net/documents/recording-historic-graffiti-advice-and-guidance-1-recording-historic-graffiti.html>

Chisholme, David. 2008. "Historic Barns of America." <https://web.archive.org/web/20160323132438/http://www.preservemassbarns.org/folklore.html> Accessed 04-21-2021.

Doherty, Jim. 1989. "A Barn is More Than a Building. It is a Shrine to Our Agrarian Past." *Smithsonian Magazine,* August.

Donmoyer, Patrick J. 2019. *Hex Signs: Sacred & Celestial Symbolism in Pennsylvania Dutch Barn Stars.* A Collaborative Exhibition with Glencairn Museum. Kutztown, PA: Pennsylvania German Cultural Heritage Center, Kutztown University.

— —. 2014 "The Concealment of Blessings in Pennsylvania Barns," *Historical Archaeology: Manifestations of Magic: The Archaeology and Material Culture of Folk Religion.* Montclair, NJ: Society for Historical Archaeology, XL(6):179-195.

— —. 2013. *Hex Signs: Myth & Meaning in Pennsylvania Dutch Barn Stars.* Kutztown, PA: Pennsylvania German Cultural Heritage Center, Kutztown University.

Donmoyer, Patrick J. & Jeffrey Marshall, et al. 2016. *Over the Blue Mountain: Barn Tour Book of the Historic Barn and Farm Foundation of Pennsylvania.* Kutztown, PA: Pennsylvania German Cultural Heritage Center, Kutztown University.

Dornbusch, Charles H. & John K. Heyl. 1958. *Pennsylvania German Barns.* Yearbook of the Pennsylvania German Folklore Society, XXI. Allentown, PA: Schlechter's Printing.

Ensminger, Robert F. 2003. *The Pennsylvania Barn: Its Origin, Evolution, and Distribution in North America.* Baltimore, MD: Johns Hopkins University Press.

— —. 1980-1981. "A Search for the Origin of the Pennsylvania Barn." *Pennsylvania Folklife* XXX(2).

— —. 1983. "A Comparative Study of Pennsylvania and Wisconsin Forebay Barns." *Pennsylvania Folklife* XXXII(3).

Fooks, David. 2004. "The History of Pennsylvania's Barn Stars and Hex Signs" *Material Culture*. Fall XXXVI(20): 1-7.

Fooks, David. 2003. "The History of Hex Signs." *The Pennsylvania German Review,* Fall 2003. Kutztown, PA: Pennsylvania German Cultural Heritage Center.

Fooks, David. 2002. "In Search of America's Oldest Hex Signs," *Der Reggeboge* 36(1):21–27.

Glass, Joseph W. 1986. *The Pennsylvania Culture Region: A View from the Barn.* Ann Arbor, MI: UMI Research Press.

Heffner, Lee S. & Patrick J. Donmoyer. 2020. *Painter of the Stars: The Life and Work of Milton J. Hill (1887-1972).* Kutztown, PA: Pennsylvania German Cultural Heritage Center, Kutztown University.

Lambert, Marcus Bachman. 1924. *A Dictionary of the Non-English Words of the Pennsylvania-German Dialect with an Appendix.* Lancaster, PA: The Pennsylvania German Society.

Long, Amos, Jr. 1972. *The Pennsylvania German Family Farm: A Regional Architectural and Folk Cultural Study of an American Agricultural Community.* Breinigsville, PA: Pennsylvania German Society.

Marshall, Jeffrey L. & Willis M. Rivinus. 2007. *Barns of Bucks County.* Doylestown, PA: Heritage Conservancy & The Bucks County Audubon Society.

Marshall, Jeffrey L. 2006. *Pennsylvania German Barns of Southeastern Pennsylvania.* Harleysville, PA: Mennonite Historians of Eastern Pennsylvania & Doylestown, PA: Heritage Conservancy

McIlwraith, Thomas F. 1981."The Diamond Cross: An Enigmatic Sign in the Rural Ontario Landscape" *Pioneer America.* XII:(1).

Mott, John. 1978. *Barns of New England: Surveyed by John A. Mott and Frank White for Old Sturbridge Village.* Unpublished Report.

Noble, Allen G., Hubert G. H. Willhelm, & Timothy G. Anderson. 2018. *Barns of the Midwest.* Athens: Ohio University Press.

Shoemaker, Alfred L. 1953. *Hex No!* Lancaster, PA: Pennsylvania Dutch Folklore Center, Franklin & Marshall College.

Shoemaker, Alfred L. et al. 1955. *The Pennsylvania Barn.* Lancaster, PA: Pennsylvania Dutch Folklore Center, Franklin & Marshall College.

Shuffelton, Frank. 1993. *A Mixed Race: Ethnicity in Early America.* New York: Oxford University Press.

Yoder, Don & Thomas E. Graves. 2000. *Hex Signs: Pennsylvania Dutch Barn Symbols and Their Meanings.* 2nd Ed. Mechanicsburg, PA: Stackpole Books.

INDEX